ATOMIC STRUCTURE

THE ATOM

AN ATOM IS THE SMALLEST DIVISIBLE UNIT OF ANY SUBSTANCE THAT CAN STILL BE IDENTIFIED AS THAT SUBSTANCE.

SUB-ATOMIC PARTICLES

Atoms consist of protons, neutrons and electrons. The NUCLEUS at the centre of the atom CONTAINS PROTONS AND NEUTRONS, and ELECTRONS ORBIT THE NUCLEUS.

A NEUTRAL ATOM HAS the SAME NUMBER OF PROTONS AND ELECTRONS.

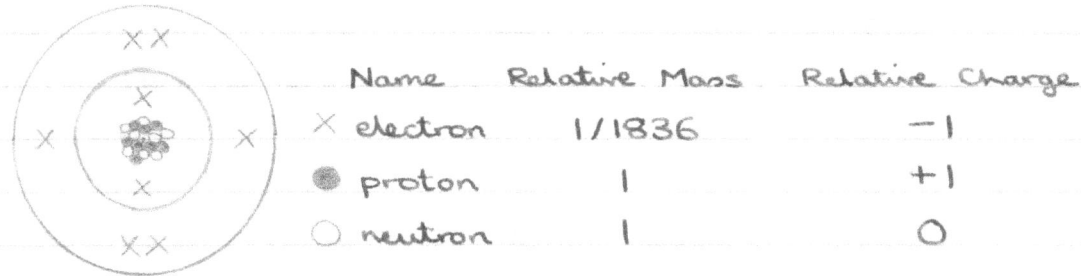

	Name	Relative Mass	Relative Charge
$\times$	electron	1/1836	-1
●	proton	1	$+1$
○	neutron	1	0

The ATOMIC NUMBER (Z) of an atom is the NUMBER OF PROTONS IN its NUCLEUS.

The MASS NUMBER (A) of an atom is the TOTAL NUMBER OF NUCLEONS (protons and neutrons) IN its NUCLEUS.

E.G. sodium

$$^{A\ 23}_{Z\ 11}Na$$

An ELEMENT is DEFINED BY the NUMBER OF

PROTONS IN the NUCLEI OF ITS ATOMS.

The NUMBER OF ELECTRONS IN an ELEMENT'S ATOMS DETERMINE its CHEMICAL PROPERTIES.

A MOLECULE CONSISTS OF TWO OR MORE ATOMS that are HELD TOGETHER BY CHEMICAL BONDS.

ISOTOPES

ISOTOPES OF a PARTICULAR ELEMENT ARE ATOMS CONTAINING the SAME NUMBER OF PROTONS BUT DIFFERENT NUMBERS OF NEUTRONS.

The ATOMS OF the MAJORITY OF ELEMENTS EXIST AS a MIXTURE OF ISOTOPES.

RELATIVE ISOTOPIC MASS

THE RELATIVE ISOTOPIC MASS IS THE MASS OF AN ATOM OF A SINGLE ISOTOPE RELATIVE TO ONE TWELFTH OF THE MASS OF A CARBON-12 ATOM.

The MASSES OF ISOTOPES are MEASURED ON a RELATIVE SCALE (scale on which an atom of carbon-12 weighs exactly 12 units) BECAUSE the MASSES OF PROTONS AND NEUTRONS VARY SLIGHTLY even from one nucleus to another. RELATIVE ISOTOPIC MASSES ARE NUMERICALLY ROUGHLY EQUIVALENT TO the MASS NUMBER OF the ATOM.

RELATIVE ATOMIC MASS (RAM)

THE RELATIVE ATOMIC MASS IS THE WEIGHTED AVERAGE OF THE MASSES OF THE ISOTOPES REL-

ATIVE TO ONE TWELFTH OF THE MASS OF A CARB-
ON-12 ATOM.

MASS SPECTROMETRY

The MASS SPECTROMETER is USED FOR FINDING
the RELATIVE ISOTOPIC MASSES AND NATURAL ABU-
NDANCES OF the ISOTOPES IN an ELEMENT. These can
then be used to CALCULATE the RELATIVE ATOMIC
MASS.

Diagram of a mass spectrometer

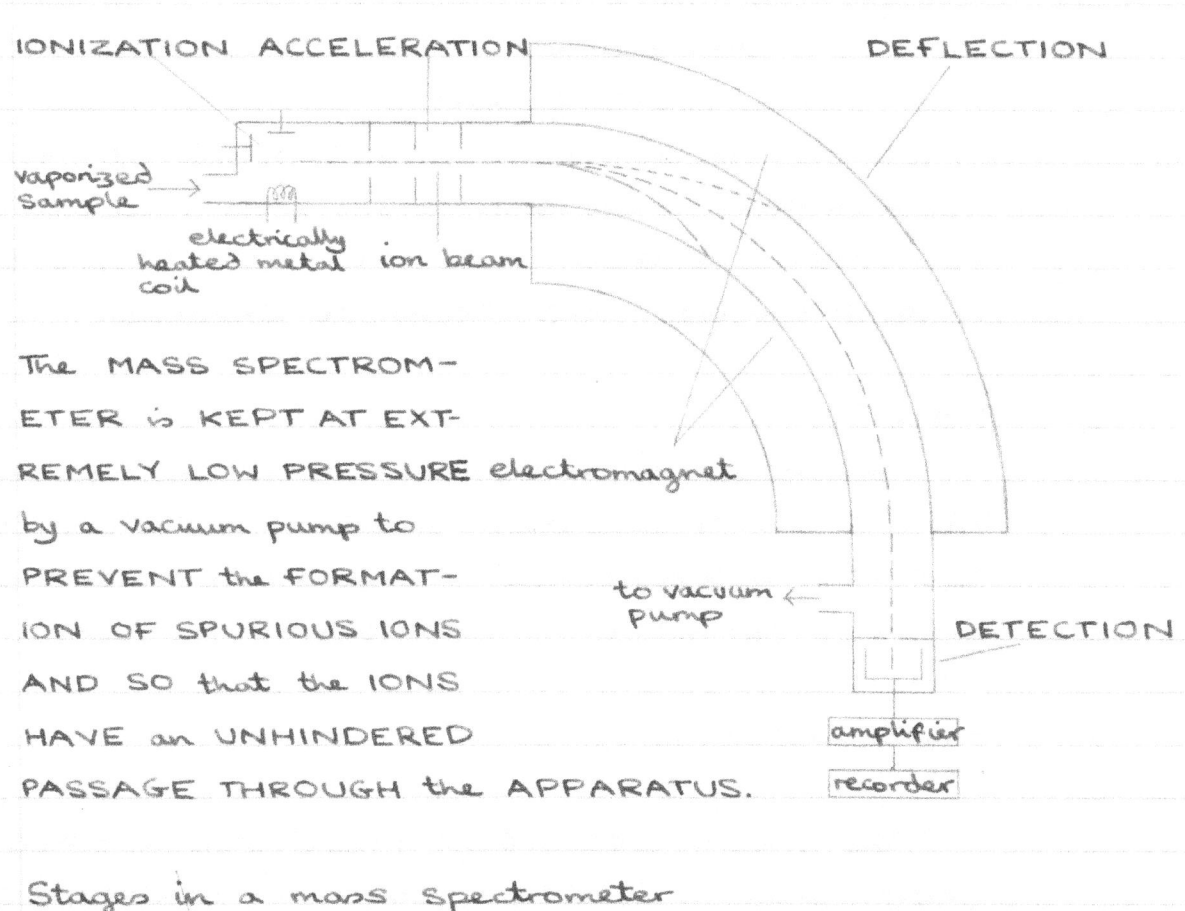

IONIZATION ACCELERATION

DEFLECTION

vaporized
Sample

electrically
heated metal   ion beam
coil

The MASS SPECTROM-
ETER is KEPT AT EXT-
REMELY LOW PRESSURE electromagnet
by a vacuum pump to
PREVENT the FORMAT-
ION OF SPURIOUS IONS   to vacuum
                               pump
AND SO that the IONS
HAVE an UNHINDERED   DETECTION
PASSAGE THROUGH the APPARATUS.   amplifier

                                     recorder

Stages in a mass spectrometer

Ionization

The VAPORIZED SAMPLE IS POSITIVELY IONIZED

BY BOMBARDMENT WITH ELECTRONS, given off by an electrically heated metal coil. The IONS produced are REPELLED INTO THE REST OF THE MACHINE by a slightly positively charged metal plate (ION REPELLER).

Acceleration

The POSITIVE IONS ARE ACCELERATED INTO a FINELY FOCUSED BEAM BY ATTRACTION TO NEG-ATIVELY CHARGED PLATES WITH CENTRAL SLITS.

Deflection

A VARIABLE MAGNETIC FIELD DEFLECTS the IONS BY DIFFERENT AMOUNTS DEPENDING ON their MASS/CHARGE RATIO (m/z or m/e).

IONS OF LOW MASS are DEFLECTED MORE TH-AN HEAVIER IONS, AND IONS WITH TWO OR MORE POSITIVE CHARGES are DEFLECTED MORE THAN IONS WITH only ONE POSITIVE CHARGE.

Detection

The POSITIVELY CHARGED IONS are NEUTRALIZED BY a FLOW OF ELECTRONS, which is DETECTED AS an ELECTRIC CURRENT that can be AMPLIFIED AND RECO-RDED. The DETECTOR also MEASURES the NUMBER OF IONS

MASS SPECTRA

The CHART RECORDER PRODUCES a MASS SPEC-TRUM, which is a STICK DIAGRAM SHOWING the RE-LATIVE CURRENT PRODUCED BY IONS OF VARYING MASS/CHARGE RATIO. The HEIGHT OF EACH PEAK is PR-OPORTIONAL to the PERCENTAGE OF THAT ISOTOPE PRESENT.

E.G. MASS SPECTRUM OF GERMANIUM

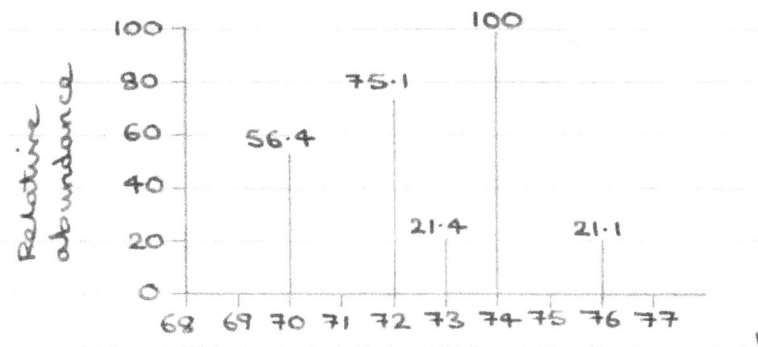

Total No. of atoms = 56.4 + 75.1 + 21.4 + 100 + 21.1 = 274

Total mass of 274 atoms = (56.4 × 70) + (75.1 × 72) + (21.4 × 73)
                        + (100 × 74) + (21.1 × 76) = 19921

Average mass of 1 atom (RAM) = 19921 / 274 = 72.7

E.G. MASS SPECTRUM OF MAGNESIUM

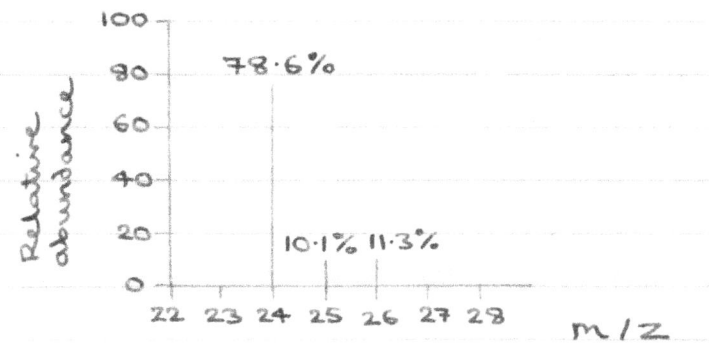

Assume there are 100 typical atoms

Total mass = (78.6 × 24) + (10.1 × 24) + (11.3 × 26) = 2432.7

RAM = 2432.7 / 100 = 24.3 (3 s.f.)

E.G.   MASS SPECTRUM OF CHLORINE

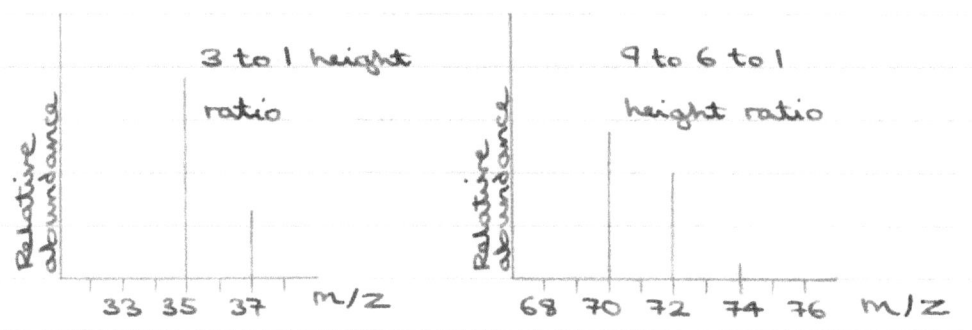

CHLORINE EXISTS AS Cl_2 MOLECULES, NOT INDIV-IDUAL ATOMS. Its atoms consist of 75% ^{35}Cl isoto-pes and 25% ^{37}Cl isotopes.

The MANY DIFFERENT PEAKS IN the MASS SPE-CTRUM are DUE TO FRAGMENTATION OF SOME OF THE MOLECULES, GIVING a CHLORINE ATOM AND a Cl^+ ION :

$$Cl_2^+ \longrightarrow Cl + Cl^+$$

UNLESS the CHLORINE ATOMS formed ARE then IONIZED, THEY WILL PASS THROUGH the MACHINE UN-DETECTED. The Cl^+ IONS WILL BE DETECTED, AND GIVE LINES FOR $^{35}_{17}Cl^+$ and $^{37}_{17}Cl^+$ IN the RATIO OF 3:1 RESPECTIVELY. The RAM of chlorine can therefore be calculated :

$$3/4 \times 35 + 1/4 \times 37 = 35.5$$

The UNFRAGMENTED Cl_2^+ IONS will also GIVE LINES IN the MASS SPECTRUM, the MASSES AND RELATI-VE HEIGHTS OF WHICH DEPENDING ON the COMBINA-TION of ^{35}Cl and ^{37}Cl ISOTOPES PRESENT :

	m/z	height ratio
$(^{35}Cl-^{35}Cl)^+$	70	$(3/4 \times 3/4) = 9/16$
$(^{35}Cl-^{37}Cl)^+$	72	$(3/4 \times 1/4) = 6/16$
$(^{37}Cl-^{37}Cl)^+$	74	$(1/4 \times 1/4) = 1/16$

ELECTRONIC STRUCTURE OF THE ATOM

ENERGY LEVELS / SHELLS

The ELECTRONS in an ATOM are arranged into SHELLS or ENERGY LEVELS. The energy levels are numbered $n=1$, $n=2$, $n=3$ etc. These numbers are called the PRINCIPAL QUANTUM NUMBERS. The LEVEL OF LOWEST ENERGY is given the NUMBER 1, the NEXT LOWEST 2, and so on.

Each ENERGY LEVEL can HOLD UP TO a certain MAXIMUM NUMBER OF ELECTRONS.

n	Shell	Max. number of electrons
1	K or 1ST	2
2	L or 2ND	8
3	M or 3RD	18
4	N or 4TH	32
5	O or 5TH	50

The MAXIMUM NUMBER OF ELECTRONS a SHELL CAN HOLD is GIVEN BY $2n^2$ (n can be from 1-7).

SUBSHELLS AND ORBITALS

The PRINCIPAL ENERGY LEVELS of electrons in atoms can be ARRANGED INTO SUBSETS which are LABELLED s, p, d and f and CALLED SUBSHELLS. (The letters s, p, d and f stem from the words sharp, principal, diffuse and fine used to describe lines in spectra).

SUBSHELLS are further DIVIDED INTO ORBITALS. AN ORBITAL IS A REGION WHERE THE ELECTRON IS MOST LIKELY TO BE FOUND. Each ORBITAL can HOLD EITHER ONE OR a MAXIMUM OF TWO ELECTRONS.

The table below shows the arrangement of electrons in the orbitals of each subshell.

Subshell	orbitals	electrons
s	1	2
p	3	6
d	5	10
f	7	14

The SHAPES of s and p ORBITALS are SHOWN BELOW. ALL the ORBITALS are 3-DIMENSIONAL. The S ORBITALS are SPHERICAL in shape; the p ORBITALS approximately 'DUMB-BELL' SHAPED with the nucleus located between the two halves of the 'dumb-bell'. The THREE P ORBITALS are IDENTICAL except for their AXES OF SYMMETRY, which are MUTUALLY AT RIGHT ANGLES.

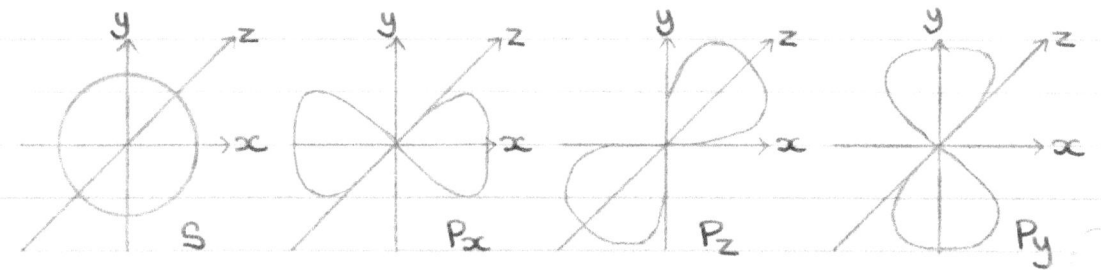

NOTATION

The ELECTRONIC STRUCTURE of an ATOM can be DESCRIBED IN TERMS of its SUBSHELLS OCCUPIED BY ELECTRONS. A NUMBER (1,2,3 etc) is USED TO DENOTE the PRINCIPAL ENERGY LEVEL, a LETTER (s, p, d or f) TO DENOTE the SUBSHELL and a SUPERSCRIPT to INDICATE the NUMBER OF ELECTRONS in the SUBSHELL.

For example, a 1s subshell holding 2 electrons would be written as $1s^2$ and read as 'ONE S TWO'.

This could also be represented as ARROWS IN A

BOX:

$$1s$$

$$\boxed{\uparrow\downarrow}$$

The BOX REPRESENTS a 1s ORBITAL and the ARROWS REPRESENT ELECTRONS. The OPPOSITE SPINS of the PAIRED ELECTRONS are SHOWN by DRAWING ARROWS IN OPPOSITE DIRECTIONS.

ORDER OF FILLING ORBITALS

ELECTRONS ALWAYS FILL THE LOWEST ENERGY ORBITALS FIRST AND THE OTHER ORBITALS IN ORDER OF ASCENDING ENERGY. (The 1s ORBITAL has the LOWEST ENERGY, the 2s THE NEXT LOWEST etc).

Where there are SEVERAL ORBITALS of EQUAL ENERGY, ELECTRONS FILL the ORBITALS SINGLY AS FAR AS POSSIBLE.

The ORDER of FILLING ORBITALS is
1s 2s 2p 3s 3p 4s 3d 4p

When the NUMBER OF ELECTRONS in EACH SUBSHELL has been WORKED OUT, the SUBSHELLS with the SAME PRINCIPAL QUANTUM NUMBER must be WRITTEN ADJACENT TO EACH OTHER. For example $1s^2 2s^2 2p^6 3s^2 3p^6 4s^2 3d^3$ becomes $1s^2 2s^2 2p^6 3s^2 3p^6 3d^3 4s^2$.

ELECTRONIC CONFIGURATIONS FROM H TO Kr

The electronic configurations of atoms of the elements hydrogen – Krypton are shown on the following pages.

ELEMENT	SYMBOL	s, p, d, f electron notation	electrons-in-boxes notation							
			1s	2s	2p	3s	3p	3d	4s	4p
HYDROGEN	H	$1s^1$	↑							
HELIUM	He	$1s^2$	↑↓							
LITHIUM	Li	$1s^2 2s^1$	↑↓	↑						
BERYLLIUM	Be	$1s^2 2s^2$	↑↓	↑↓						
BORON	B	$1s^2 2s^2 2p^1$	↑↓	↑↓	↑					
CARBON	C	$1s^2 2s^2 2p^2$	↑↓	↑↓	↑ ↑					
NITROGEN	N	$1s^2 2s^2 2p^3$	↑↓	↑↓	↑ ↑ ↑					
OXYGEN	O	$1s^2 2s^2 2p^4$	↑↓	↑↓	↑↓ ↑ ↑					
FLUORINE	F	$1s^2 2s^2 2p^5$	↑↓	↑↓	↑↓ ↑↓ ↑					
NEON	Ne	$1s^2 2s^2 2p^6$	↑↓	↑↓	↑↓ ↑↓ ↑↓					
SODIUM	Na	$[Ne]3s^1$	↑↓	↑↓	↑↓ ↑↓ ↑↓	↑				
MAGNESIUM	Mg	$[Ne]3s^2$	↑↓	↑↓	↑↓ ↑↓ ↑↓	↑↓				
ALUMINIUM	Al	$[Ne]3s^2 3p^1$	↑↓	↑↓	↑↓ ↑↓ ↑↓	↑↓	↑			
SILICON	Si	$[Ne]3s^2 3p^2$	↑↓	↑↓	↑↓ ↑↓ ↑↓	↑↓	↑ ↑			
PHOSPHOROUS	P	$[Ne]3s^2 3p^3$	↑↓	↑↓	↑↓ ↑↓ ↑↓	↑↓	↑ ↑ ↑			
SILICON	Si	$[Ne]3s^2 3p^2$	↑↓	↑↓	↑↓ ↑↓ ↑↓	↑↓	↑ ↑			
SULPHUR	S	$[Ne]3s^2 3p^4$	↑↓	↑↓	↑↓ ↑↓ ↑↓	↑↓	↑↓ ↑ ↑			
CHLORINE	Cl	$[Ne]3s^2 3p^5$	↑↓	↑↓	↑↓ ↑↓ ↑↓	↑↓	↑↓ ↑↓ ↑			
ARGON	Ar	$[Ne]3s^2 3p^6$	↑↓	↑↓	↑↓ ↑↓ ↑↓	↑↓	↑↓ ↑↓ ↑↓			

ELECTRONIC CONFIGURATIONS OF ATOMS OF
THE ELEMENTS H – Ar

ELEMENT	SYMBOL	s,p,d,f electron notation	1s	2s	2p	3s	3p	3d	4s	4p
POTASSIUM	K	$[Ar]4s^1$	↑↓	↑↓	↑↓ ↑↓ ↑↓	↑↓	↑↓ ↑↓ ↑↓		↑	
CALCIUM	Ca	$[Ar]4s^2$	↑↓	↑↓	↑↓ ↑↓ ↑↓	↑↓	↑↓ ↑↓ ↑↓		↑↓	
SCANDIUM	Sc	$[Ar]3d^1 4s^2$	↑↓	↑↓	↑↓ ↑↓ ↑↓	↑↓	↑↓ ↑↓ ↑↓	↑	↑↓	
TITANIUM	Ti	$[Ar]3d^2 4s^2$	↑↓	↑↓	↑↓ ↑↓ ↑↓	↑↓	↑↓ ↑↓ ↑↓	↑ ↑	↑↓	
VANADIUM	V	$[Ar]3d^3 4s^2$	↑↓	↑↓	↑↓ ↑↓ ↑↓	↑↓	↑↓ ↑↓ ↑↓	↑ ↑ ↑	↑↓	
CHROMIUM	Cr	$[Ar]3d^5 4s^1$	↑↓	↑↓	↑↓ ↑↓ ↑↓	↑↓	↑↓ ↑↓ ↑↓	↑ ↑ ↑ ↑ ↑	↑	
MANGANESE	Mn	$[Ar]3d^5 4s^2$	↑↓	↑↓	↑↓ ↑↓ ↑↓	↑↓	↑↓ ↑↓ ↑↓	↑ ↑ ↑ ↑ ↑	↑↓	
IRON	Fe	$[Ar]3d^6 4s^2$	↑↓	↑↓	↑↓ ↑↓ ↑↓	↑↓	↑↓ ↑↓ ↑↓	↑↓ ↑ ↑ ↑ ↑	↑↓	
COBALT	Co	$[Ar]3d^7 4s^2$	↑↓	↑↓	↑↓ ↑↓ ↑↓	↑↓	↑↓ ↑↓ ↑↓	↑↓ ↑↓ ↑ ↑ ↑	↑↓	
NICKEL	Ni	$[Ar]3d^8 4s^2$	↑↓	↑↓	↑↓ ↑↓ ↑↓	↑↓	↑↓ ↑↓ ↑↓	↑↓ ↑↓ ↑↓ ↑ ↑	↑↓	
COPPER	Cu	$[Ar]3d^{10} 4s^1$	↑↓	↑↓	↑↓ ↑↓ ↑↓	↑↓	↑↓ ↑↓ ↑↓	↑↓ ↑↓ ↑↓ ↑↓ ↑↓	↑	
ZINC	Zn	$[Ar]3d^{10} 4s^2$	↑↓	↑↓	↑↓ ↑↓ ↑↓	↑↓	↑↓ ↑↓ ↑↓	↑↓ ↑↓ ↑↓ ↑↓ ↑↓	↑↓	
GALLIUM	Ga	$[Ar]3d^{10} 4s^2 4p^1$	↑↓	↑↓	↑↓ ↑↓ ↑↓	↑↓	↑↓ ↑↓ ↑↓	↑↓ ↑↓ ↑↓ ↑↓ ↑↓	↑↓	↑
GERMANIUM	Ge	$[Ar]3d^{10} 4s^2 4p^2$	↑↓	↑↓	↑↓ ↑↓ ↑↓	↑↓	↑↓ ↑↓ ↑↓	↑↓ ↑↓ ↑↓ ↑↓ ↑↓	↑↓	↑ ↑
ARSENIC	As	$[Ar]3d^{10} 4s^2 4p^3$	↑↓	↑↓	↑↓ ↑↓ ↑↓	↑↓	↑↓ ↑↓ ↑↓	↑↓ ↑↓ ↑↓ ↑↓ ↑↓	↑↓	↑ ↑ ↑
SELENIUM	Se	$[Ar]3d^{10} 4s^2 4p^4$	↑↓	↑↓	↑↓ ↑↓ ↑↓	↑↓	↑↓ ↑↓ ↑↓	↑↓ ↑↓ ↑↓ ↑↓ ↑↓	↑↓	↑↓ ↑ ↑
BROMINE	Br	$[Ar]3d^{10} 4s^2 4p^5$	↑↓	↑↓	↑↓ ↑↓ ↑↓	↑↓	↑↓ ↑↓ ↑↓	↑↓ ↑↓ ↑↓ ↑↓ ↑↓	↑↓	↑↓ ↑↓ ↑
KRYPTON	Kr	$[Ar]3d^{10} 4s^2 4p^6$	↑↓	↑↓	↑↓ ↑↓ ↑↓	↑↓	↑↓ ↑↓ ↑↓	↑↓ ↑↓ ↑↓ ↑↓ ↑↓	↑↓	↑↓ ↑↓ ↑↓

electrons-in-boxes notation

ELECTRONIC CONFIGURATIONS OF ATOMS
OF THE ELEMENTS K–Kr

In order to SIMPLIFY WRITING ELECTRONIC CONFIGURATIONS in s, p, d, f notation, the INNER ELECTRON CONFIGURATION OF an ELEMENT is WRITTEN AS the SYMBOL FOR the PREVIOUS NOBLE GAS, AND the REMAINING ORBITALS are WRITTEN AS BEFORE.

E.G. Silicon is written as $[Ne] 3s^2 3p^2$ rather than $1s^2 2s^2 2p^6 3s^2 3p^2$

Notice the UNEXPECTED ELECTRON STRUCTURES for CHROMIUM and COPPER.

The arrangement of electrons in CHROMIUM is $[Ar] 3d^5 4s^1$ NOT $[Ar] 3d^4 4s^2$ as might have been expected. ONE OF THE S ELECTRONS MOVES TO the d SHELL, LEAVING HALF-FILLED 3d AND 4S SUBSHELLS, because this ARRANGEMENT has a LOWER ENERGY LEVEL THAN the EXPECTED ONE. The EXTRA STABILITY OF a HALF-FILLED SUBSHELL is THOUGHT TO RESULT FROM ITS SYMMETRY, AS the OCCUPATION OF EACH ORBITAL BY ONE ELECTRON PRODUCES an EQUAL DISTRIBUTION OF CHARGE AROUND an ATOM

Similarly, the arrangement of electrons in COPPER is $[Ar] 3d^{10} 4s^1$ RATHER THAN $[Ar] 3d^9 4s^2$. In this case, the FIRST ARRANGEMENT, with a FILLED 3d SUBSHELL and a HALF-FILLED 4S SUBSHELL, is MORE STABLE THAN the LATTER ARRANGEMENT, which is again ASSOCIATED WITH the SYMMETRY of a FULL d SUBSHELL.

ELECTRONIC CONFIGURATIONS OF MONATOMIC IONS

Certain MONATOMIC IONS FORM WHEN ATOMS GAIN OR LOSE ELECTRONS TO ACHIEVE STABLE ELECTR-

ON CONFIGURATIONS FOR their HIGHEST ENERGY ELECTRONS.

NON-METALLIC ELEMENTS GAIN ELECTRONS to form ANIONS and METALLIC ELEMENTS LOSE ELECTRONS to form CATIONS.

WHEN WRITING the ELECTRONIC CONFIGURATIONS OF IONS, FIRST WRITE the CONFIGURATION FOR the NEUTRAL ATOM, THEN ADD OR SUBTRACT 1, 2 OR 3 ELECTRONS (starting from the highest energy level orbitals) to give OUTER ELECTRON CONFIGURATIONS of $1s^2$, $ns^2 np^6$, nd^{10} or $nd^{10}(n+1)s^2$. (Atoms of elements which cannot reach these stable configurations form less easily predicted charges, which have to be memorized).

The electronic configurations of some atoms and ions are shown below.

Atom	Electronic Configuration	Ion	Electronic Configuration
Cl	$[Ne] 3s^2 3p^5$	Cl^-	$[Ne] 3s^2 3p^6$
Li	$1s^2 2s^1$	Li^+	$1s^2$
Mg	$[Ne] 3s^2$	Mg^{2+}	$[Ne]$
Zn	$[Ar] 3d^{10} 4s^2$	Zn^{2+}	$[Ar] 3d^{10}$
Sc	$[Ar] 3d^1 4s^2$	Sc^{3+}	$[Ar]$
S	$[Ne] 3s^2 3p^4$	S^{2-}	$[Ne] 3s^2 3p^6$
O	$1s^2 2s^2 2p^4$	O^{2-}	$[Ne]$
Na	$[Ne] 3s^1$	Na^+	$[Ne]$
Ca	$[Ne] 3s^2 3p^6 4s^2$	Ca^{2+}	$[Ne] 3s^2 3p^6$

IONIZATION ENERGY

IONIZATION ENERGY is a MEASURE OF THE EASE WITH WHICH ELECTRONS ARE LOST IN the FORMATION OF POSITIVE IONS.

THE FIRST IONIZATION ENERGY IS THE ENERGY REQUIRED TO REMOVE ONE ELECTRON FROM EACH ATOM IN

A MOLE OF GASEOUS ATOMS TO PRODUCE ONE MOLE OF GASEOUS 1+ IONS.

$$X(g) \longrightarrow X^+(g) + e^-$$

Successive ionization energies are for the removal of a second and subsequent electrons from a gaseous ion.

THE SECOND IONIZATION ENERGY IS THE ENERGY REQUIRED TO REMOVE ONE ELECTRON FROM EACH ION IN A MOLE OF GASEOUS 1+ IONS TO PRODUCE ONE MOLE OF GASEOUS 2+ IONS.

$$X^+(g) \longrightarrow X^{2+}(g) + e^-$$

Ionization energies are measured in KILOJOULES PER MOLE ($KJ mol^{-1}$). They have POSITIVE VALUES as ENERGY is REQUIRED TO REMOVE AN ELECTRON FROM the ATTRACTION OF a POSITIVE NUCLEUS (IONIZATION is an ENDOTHERMIC PROCESS). The STRENGTH OF the ATTRACTION DETERMINES the MAGNITUDE of the IONIZATION ENERGY and depends upon:

• The size of the nuclear charge

INCREASING the NUMBER OF PROTONS IN the NUCLEUS WILL INCREASE ITS ATTRACTION FOR the OUTERMOST ELECTRON, INCREASING IONIZATION ENERGY.

• The screening (shielding) effect of inner electrons

OUTER SHELL ELECTRONS are SCREENED FROM the ATTRACTION OF the NUCLEUS BY the REPELLING EFFECT OF INNER ELECTRONS. This REDUCES the OVERALL ATTRACTIVE FORCE FELT BY OUTER ELECTRONS, REDUCING IONIZATION ENERGY.

• The distance of the outermost electron from the nucleus

   The FURTHER the ELECTRON is FROM THE NUCLEUS, the LESS THE ATTRACTION, DECREASING IONIZATION ENERGY.

SUCCESSIVE IONIZATION ENERGIES

   AFTER REMOVING the FIRST ELECTRON FROM AN ATOM, REMOVING SUBSEQUENT ELECTRONS becomes PROGRESSIVELY MORE DIFFICULT, therefore SUCCESSIVE IONIZATION ENERGIES INCREASE. This is because AS ELECTRONS are REMOVED, REMAINING ELECTRONS are HELD MORE FIRMLY BY the POSITIVELY CHARGED ION LEFT BEHIND.

   SUDDEN INCREASES, or 'JUMPS' in IONIZATION ENERGY as CONSECUTIVE ELECTRONS are REMOVED, PROVIDE EVIDENCE for the EXISTENCE of ENERGY LEVELS or QUANTUM SHELLS. The SECOND ELECTRON is IN A NEW ENERGY LEVEL or SHELL, CLOSER to the NUCLEUS and WITH FEWER SCREENING ELECTRONS, THEREFORE it is MORE DIFFICULT to REMOVE.

E.G. Consider the graph of the logarithm to base ten (lg)
      for the successive ionization energies of sodium:

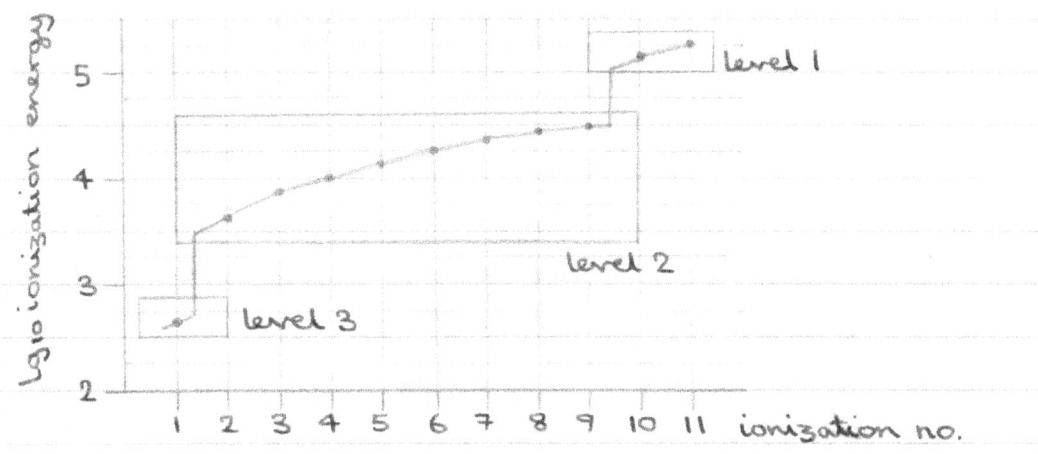

Shown on the graph are the number of electrons in

each of the energy levels of a sodium atom.

USING IONIZATION ENERGY DATA, the ELECTRONIC STRUCTURE of an ATOM can be WRITTEN in TERMS OF its ENERGY LEVELS, for example sodium is written as 2,8,1.

The DIFFERENT ENERGY LEVELS marked on the graph are RECOGNIZED by a 'JUMP' IN ENERGY AS the 2ND AND 10TH ELECTRONS are REMOVED, INDICATING the BEGINNING OF the 2ND AND 1st ELECTRON SHELLS RESPECTIVELY.

The FIRST 'JUMP' IN ENERGY BETWEEN TWO SUCCESSIVE IONIZATION ENERGIES OF AN ELEMENT can be USED TO WORK OUT WHICH GROUP the ELEMENT IS IN.

AS A 'JUMP' INDICATES the START OF a NEW ELEC-TRON SHELL, the NUMBER OF IONIZATIONS UP TO the FIRST 'JUMP' CORRESPONDS TO the NUMBER OF OUTER ELECTR-ONS, which IDENTIFIES the GROUP that the ELEMENT BELONGS TO.

E.G. The successive ionization energies of an element in KJ mol⁻¹ are:

1ST	2ND	3RD	4TH	5TH	6TH
1012	1903	2912	4957	6274	21 269

There is a STEADY INCREASE in ENERGY NEEDED to REMOVE SUCCESSIVE ELECTRONS, then a 'JUMP' AFTER the FIFTH ELECTRON is REMOVED, CORRESPONDING TO the REMOVAL OF an ELECTRON FROM a NEW SHELL. AS there are FIVE OUTER ELECTRONS TO REMOVE, the ELEMENT IS IN GROUP 5.

FIRST IONIZATION ENERGIES

The graph below shows the FIRST IONIZATION ENERGIES OF THE FIRST 36 ELEMENTS, plotted against atomic number.

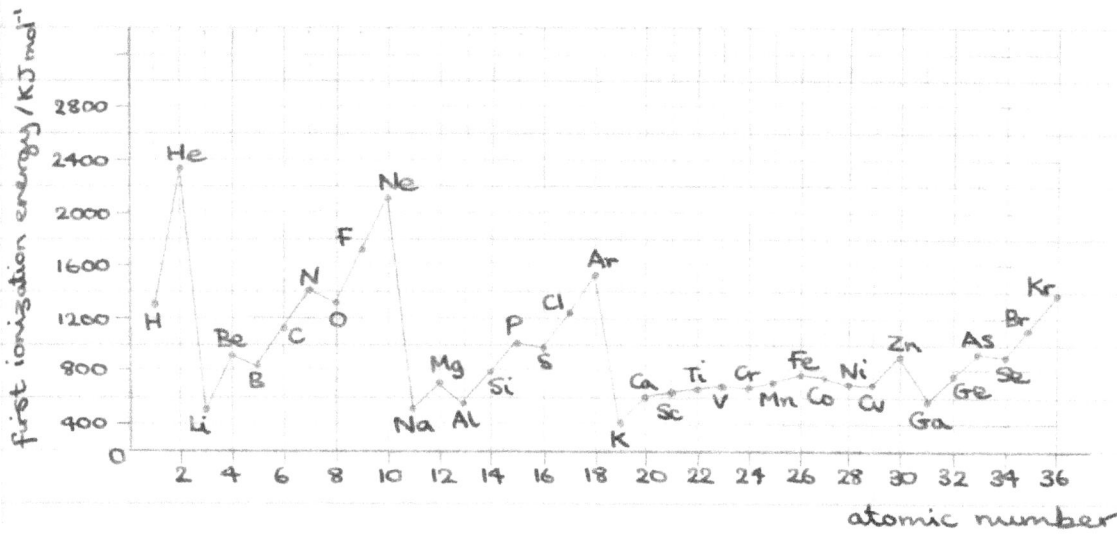

The POINTS on the SECTION OF GRAPH BETWEEN ONE NOBLE GAS AND THE NEXT CORRESPOND TO the FILLING OF A SHELL WITH ELECTRONS. These SECTIONS can be DIVIDED INTO SUB-SECTIONS, CONTAINING EITHER 2, 6 OR 10 POINTS, which CORRESPOND TO, and PROVIDE EVIDENCE FOR, the DIFFERENT SUBSHELLS OF ELECTRONS.

TRENDS IN FIRST IONIZATION ENERGY ACROSS A PERIOD

Moving FROM LEFT TO RIGHT ACROSS A PERIOD, there is an INCREASE in NUCLEAR CHARGE yet NO INCREASE in SCREENING EFFECT, as ADDITIONAL ELECTRONS are being ADDED TO the SAME OUTER SHELL. Therefore the NUCLEUS ATTRACTS the OUTER ELECTRONS MORE STRONGLY and PULLS THEM CLOSER TO IT. GENERALLY, this results in an INCREASE IN FIRST IONIZATION ENERGY ACROSS A PERIOD. However, there are DISCONTINUITIES in this TREND, which can be EXPLAINED IN TERMS OF the ELECTRONIC CONFIGURATIONS

OF the ELEMENTS CONCERNED.

In PERIOD 2 (Li to Ne), the FIRST IONIZATION ENERGY of BORON DIFFERS FROM the EXPECTED VALUE and is LOWER THAN that of BERYLLIUM. The electronic configuration of boron is $1s^2 2s^2 2p^1$ whereas that of beryllium is $1s^2 2s^2$. The ENERGY REQUIRED to REMOVE BORON'S OUTER p ELECTRON is LESS THAN THAT NEEDED to REMOVE A 2s ELECTRON FROM BERYLLIUM BECAUSE:

* the 2p ORBITAL is SLIGHTLY SHIELDED BY the FULL 2S ORBITAL.
* the 2p ORBITAL is FURTHER FROM the NUCLEUS THAN the 2S ORBITAL.
* a STABLE AND SYMMETRICAL ELECTRON CONFIGURATION is FORMED AFTER ELECTRON LOSS, known as FULL S SUBSHELL STABILITY.

Also in PERIOD 2, the FIRST IONIZATION ENERGY of OXYGEN is UNEXPECTEDLY LESS THAN that of NITROGEN. The electronic configurations of oxygen and nitrogen are $1s^2 2s^2 2p^4$ and $1s^2 2s^2 2p^3$ respectively. The ENERGY REQUIRED to REMOVE OXYGEN'S OUTER P ELECTRON is LESS THAN THAT NEEDED to REMOVE OXYGEN'S OUTER p ELECTRON is LESS THAN THAT NEEDED to REMOVE NITROGEN'S BECAUSE:

* the PAIRED ELECTRONS IN the FIRST p ORBITAL in OXY-GEN REPEL EACH OTHER.
* a STABLE AND SYMMETRICAL ELECTRON CONFIGURATION is FORMED AFTER ELECTRON LOSS, known as HALF-FULL P SUBSHELL STABILITY.

TWO SIMILAR DISCONTINUITIES in the TREND OF INCREAS-ING FIRST IONIZATION ENERGY OCCUR in PERIOD 3 (Na to Ar),

WHERE the FIRST IONIZATION ENERGIES of ALUMINIUM AND SULPHUR are LESS THAN those of MAGNESIUM AND PHOSPHOROUS RESPECTIVELY.

ALUMINIUM has a LOWER FIRST IONIZATION ENERGY VALUE THAN MAGNESIUM FOR the SAME REASON that BORON HAS a LOWER VALUE THAN BERYLLIUM, EXCEPT that EVERYTHING is HAPPENING AT the 3 LEVEL RATHER THAN the 2 LEVEL.

Similarly, SULPHUR'S FIRST IONIZATION ENERGY is LESS THAN that of PHOSPHOROUS IN the SAME WAY that OXYGEN'S is LESS THAN NITROGEN'S, again APART FROM the FACT THAT the ELECTRONS are being REMOVED FROM the 3RD SHELL INSTEAD OF the 2ND.

TRENDS IN FIRST IONIZATION ENERGY IN A TRANSITION SERIES

Going FROM LEFT TO RIGHT ACROSS the SERIES OF METALS FROM SCANDIUM TO ZINC, there is an INCREASE IN NUCLEAR CHARGE AND IN the NUMBER OF INNER d SUBSHELL ELECTRONS.

HENCE the OUTER 4S ELECTRONS are SCREENED FROM the ATTRACTION OF the INCREASINGLY POSITIVE NUCLEUS BY the EXTRA 3d ELECTRONS.

Consequently, the FIRST IONIZATION ENERGIES OF the ELEMENTS FROM SCANDIUM TO ZINC GENERALLY INCREASE, BUT ONLY MARGINALLY IN COMPARISON TO the INCREASE ACROSS PERIOD 3. However, there is a LARGER THAN NORMAL INCREASE IN ENERGY BETWEEN COPPER AND ZINC.

The ELECTRONIC CONFIGURATIONS of COPPER AND ZINC are $[Ar]3d^{10}4s^1$ and $[Ar]3d^{10}4s^2$ RESPECTIVELY. FOR BOTH ELEMENTS, the OUTERMOST ELE-

CTRON is being REMOVED FROM the SAME ORB-
ITAL, WITH IDENTICAL SCREENING, BUT ZINC HAS
an EXTRA PROTON IN the NUCLEUS thus INCREASING
the ATTRACTION.

TRENDS IN FIRST IONIZATION ENERGY DOWN A GROUP

Going DOWN A GROUP, the OUTER ELECTRON
BECOMES PROGRESSIVELY FURTHER FROM the NUCL-
EUS and there is GREATER SCREENING BECAU-
SE OF the EXTRA FILLED ORBITALS. This OFFSETS
the INCREASE IN NUCLEAR CHARGE and SO FIRST
IONIZATION ENERGY DECREASES DOWN A GROUP.

E.G. For Group I the first ionization energies, in $KJmol^{-1}$,
are:

Li	519
Na	494
K	418
Rb	402
Cs	376

ELECTRON AFFINITY

THE FIRST ELECTRON AFFINITY IS THE ENERGY
RELEASED WHEN ONE ELECTRON IS ADDED TO
EACH ATOM IN A MOLE OF GASEOUS ATOMS TO
FORM ONE MOLE OF GASEOUS I- IONS.

$$X(g) + e^- \longrightarrow X^-(g)$$

THE SECOND ELECTRON AFFINITY IS THE ENERGY REQUIRED TO ADD ONE ELECTRON TO EACH ION IN A MOLE OF GASEOUS 1- IONS TO GIVE ONE MOLE OF GASEOUS 2- IONS.

$$X^-(g) + e^- \longrightarrow X^{2-}(g)$$

The FIRST ELECTRON AFFINITY is EXOTHERMIC (negative sign) BECAUSE OF the ATTRACTION BETWEEN the POSITIVE NUCLEUS AND the NEGATIVE ELECTRON, whereas the SECOND is ENDOTHERMIC (positive sign) BECAUSE OF the REPULSION BETWEEN the NEGATIVE ANION AND the NEGATIVE ELECTRON.

E.G. The first and second electron affinities of oxygen are:

$$O(g) + e^- \longrightarrow O^-(g) \quad -142 \text{ kJ mol}^{-1}$$

$$O^-(g) + e^- \longrightarrow O^{2-}(g) \quad +844 \text{ kJ mol}^{-1}$$

BONDING

A COMPOUND CONSISTS OF ATOMS OF TWO OR MORE ELEMENTS CHEMICALLY COMBINED TOGETHER IN FIXED PROPORTIONS BY MASS.

When elements form compounds, their ATOMS EITHER LOSE, GAIN OR SHARE ELECTRONS in order TO ACHIEVE STABLE (LOW ENERGY) ELECTRON CONFIGURATIONS.

For many SIMPLE COMPOUNDS OF the S AND P BLOCK ELEMENTS, these CONFIGURATIONS are the SAME AS the PREVIOUS OR NEXT NOBLE GAS in the PERIODIC TABLE.

IONIC BONDING

AN IONIC BOND IS THE ELECTROSTATIC ATTRACTION BETWEEN OPPOSITELY CHARGED IONS.

IONIC BONDING USUALLY OCCURS WHEN ONE OR MORE ELECTRONS are TRANSFERRED FROM a METAL ATOM TO a NON-METAL ATOM, forming oppositely charged ions.

The transfer of electrons from one atom to another can be shown by means of DOT AND CROSS DIAGRAMS.

The DOTS OR CROSSES REPRESENT the ELECTRONS in each shell of an atom, and the NUCLEUS at the centre is REPRESENTED BY the SYMBOL FOR the ELEMENT. IONS are shown in SQUARE BRACKETS with the CHARGE AT the TOP RIGHT-HAND CORNER.

OFTEN ONLY the OUTER ELECTRONS of an atom ARE SHOWN IN DOT AND CROSS DIAGRAMS, AS ONLY THESE ELECTRONS ARE INVOLVED IN CHEMICAL REAC-

TIONS.

E.G.   SODIUM CHLORIDE

(2,8,1)   (2,8,7)   (2,8)   (2,8,8)
Sodium   chlorine   Sodium chloride
atom   atom   ion   ion

E.G.   LITHIUM OXIDE

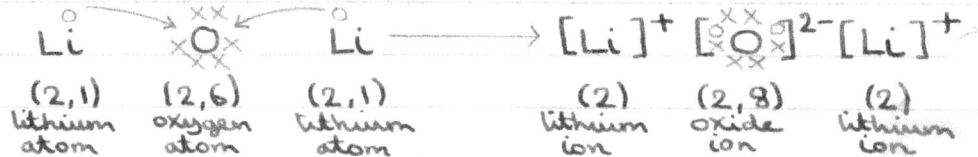

(2,1)   (2,6)   (2,1)   (2)   (2,8)   (2)
lithium   oxygen   lithium   lithium oxide   lithium
atom   atom   atom   ion   ion   ion

   DOTS AND CROSSES HELP TO SHOW WHICH VALE-
NCE ELECTRONS COME FROM WHICH ATOMS, but
ONCE A BOND IS FORMED the ELECTRONS ARE
INDISTINGUISHABLE.

COVALENT BONDING

   A COVALENT BOND IS A SHARED PAIR OF ELE-
CTRONS BETWEEN TWO ATOMS, ONE CONTRIBUTED
BY EACH ATOM.
   A COVALENT BOND FORMS BECAUSE the ATTR-
ACTION OF the NUCLEI FOR the ELECTRONS is
GREATER THAN the REPULSION BETWEEN the
NUCLEI AND BETWEEN the ELECTRONS OF the TWO
ATOMS.
   COVALENT BONDING USUALLY OCCURS WHEN
TWO NON-METAL ATOMS SHARE ELECTRONS.

E.G.   HYDROGEN

H°   ×H ⟶ H°×H

  (1)   (1)   (2)

hydrogen   hydrogen   hydrogen

atom   atom   molecule

E.G.   HYDROGEN   CHLORIDE

H°   ×Cl× ⟶ H°×Cl×

  (1)   (2,8,7)   (2) (2,8,8)

hydrogen   chlorine   hydrogen chloride

atom   atom   molecule

OTHER EXAMPLES of simple molecules are shown below as dot and cross diagrams:

WATER, H_2O AMMONIA, NH_3 METHANE, CH_4

BERYLLIUM   PHOSPHOROUS (V)   BORON

CHLORIDE, $BeCl_2$ CHLORIDE, PCl_5 TRIFLUORIDE, BF_3

MULTIPLE COVALENT BONDS

ATOMS can SHARE MORE THAN ONE PAIR OF ELECTRONS TO FORM A MULTIPLE BOND.

A DOUBLE COVALENT BOND is where TWO PAIRS OF ELECTRONS are SHARED BETWEEN ATOMS.

E.G.   OXYGEN, O_2

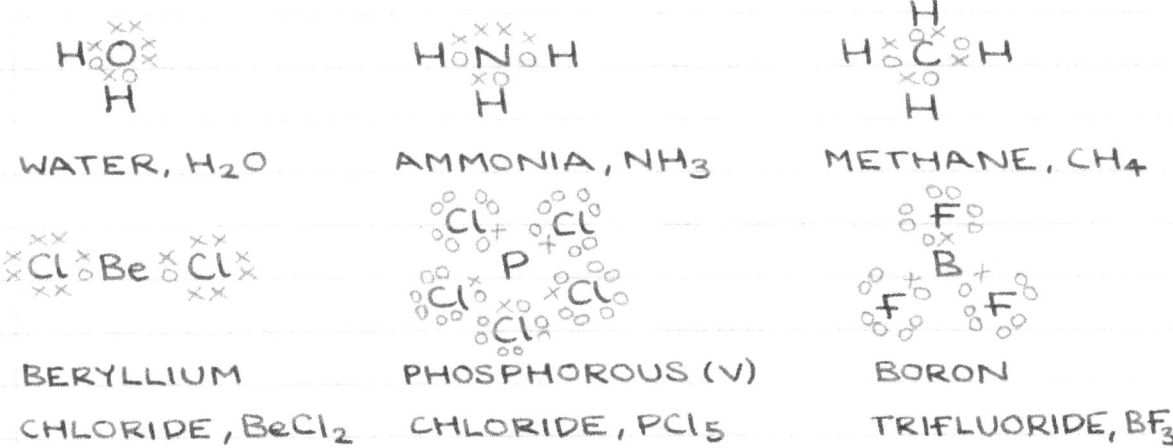

or   O = O

E.G.   CARBON DIOXIDE, CO_2

$$\overset{\circ\circ}{\underset{\circ\circ}{O}}\overset{\times\circ\times}{\underset{\circ\times\circ}{C}}\overset{\circ\circ}{\underset{\circ\circ}{O}} \qquad \text{or} \qquad O=C=O$$

A TRIPLE COVALENT BOND is where THREE
PAIRS OF ELECTRONS are SHARED BETWEEN ATOMS.

E.G.   NITROGEN, N_2

$$\overset{\circ\circ}{\underset{\circ\circ}{N}}\overset{\times\circ\times}{\underset{\times\circ\times}{N}}\overset{\times}{\underset{\times}{}} \qquad \text{or} \qquad N\equiv N$$

FOR ANY TWO GIVEN ATOMS the BOND GETS
SHORTER AND STRONGER AS IT CHANGES FROM a
SINGLE TO a DOUBLE TO a TRIPLE COVALENT
BOND.

SIGMA AND PI BONDS

A COVALENT BOND IS FORMED WHEN TWO
ATOMIC ORBITALS OVERLAP TO PRODUCE A MOLE-
CULAR ORBITAL CONTAINING THE ELECTRON
PAIR.

SIGMA (σ) BONDS

SIGMA BONDS ARE FORMED BY DIRECT (END-
TO-END) OVERLAP OF TWO ATOMIC ORBITALS.
Below are diagrams showing the FORMATION
OF SIGMA BONDS BY DIRECT OVERLAP OF TWO S,
ONE S AND ONE P, TWO p ORBITALS.

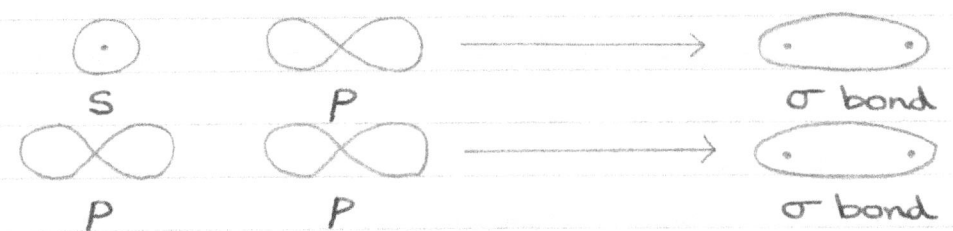

A SIGMA BOND is the ORDINARY COVALENT BOND FORMED WHEN a PAIR OF ELECTRONS is SHARED BETWEEN THE NUCLEI OF TWO ATOMS.

PI (π) BONDS

PI BONDS ARE FORMED BY SIDEWAYS OVERLAP OF TWO p ORBITALS.

Below is a diagram showing the FORMATION OF A PI BOND.

SIGMA AND PI BONDING IN DOUBLE BONDS

The TWO COVALENT BONDS IN A DOUBLE BOND are NOT IDENTICAL. ONE of the bonds is a SIGMA BOND and the OTHER is a PI BOND.

E.G. IN a MOLECULE OF ETHENE, the SIGMA BOND is FORMED BETWEEN the CARBON ATOMS and is SITUATED SYMMETRICALLY BETWEEN THEM. SIGMA BONDS are ALSO FORMED BETWEEN the CARBON AND HYDROGEN ATOMS.

The PI BOND is FORMED BY the OVERLAP OF the TWO REMAINING p ORBITALS, BOTH CONTAINING

an ELECTRON, ON EACH CARBON ATOM. AS
each p orbital has two lobes, the PI BOND has
TWO REGIONS, ONE ABOVE and ONE BELOW the
PLANE OF the MOLECULE.

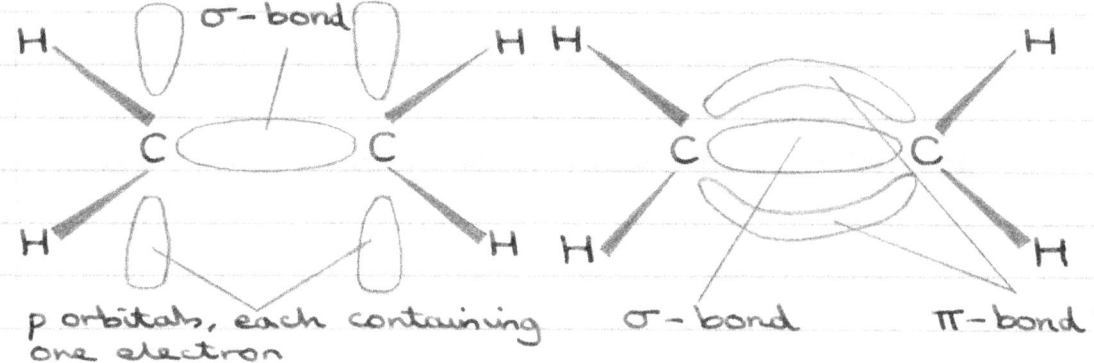

p orbitals, each containing
one electron

a) σ − bond only   b) σ−bond and π−bond

Note that the plane of the molecule is perpendic-
ular to the page.

CO − ORDINATE (DATIVE COVALENT) BONDING

        A CO−ORDINATE (DATIVE COVALENT) BOND IS
A SHARED PAIR OF ELECTRONS BETWEEN
TWO ATOMS, BOTH CONTRIBUTED BY ONE ATOM.
        CO−ORDINATE BONDING OCCURS WHEN ONE
ATOM HAS a LONE (UNSHARED) PAIR OF ELECT-
RONS WHICH ARE DONATED TO the VACANT
SHELL OF ANOTHER ATOM.

E.G.   reaction between AMMONIA AND HYDROGEN
        CHLORIDE

$$NH_3 + HCl \longrightarrow NH_4Cl$$

When AMMONIA is MIXED WITH GASEOUS

HYDROGEN CHLORIDE, WHITE CLOUDS of AMMONIUM
CHLORIDE FORM. The REACTION is DUE TO the FOR-
MATION OF a DATIVE COVALENT BOND BETWEEN
the UNSHARED PAIR OF ELECTRONS ON the N ATOM
IN NH_3 AND an H^+ ION FROM the HCl.

The COVALENT BONDS in the ammonium ion
can be REPRESENTED MORE SIMPLY AS SINGLE
LINES, and the DATIVE BOND REPRESENTED AS
AN ARROW DRAWN FROM the nitrogen (ELECTR-
ON PAIR DONOR) to the hydrogen ion (ELECTRON
PAIR ACCEPTOR):

E.G.   reaction between HYDROGEN CHLORIDE AND
       WATER

$$H_2O \; + \; HCl \longrightarrow H_3O^+ + Cl^-$$

The H_3O^+ ion formed is called the HYDROXONI-
UM ION.

ALTHOUGH the DATIVE BOND in the hydr-
oxonium ion IS SHOWN AS an ARROW, IT IS
IDENTICAL TO the OTHER COVALENT BONDS
AND CANNOT BE DISTINGUISHED ONCE FORMED.

E.G. reaction between AMMONIA AND BORON
TRIFLUORIDE

E.G. DIMERIZATION OF ALUMINIUM CHLORIDE

WHEN GASEOUS ALUMINIUM CHLORIDE is
COOLED, $AlCl_3$ MOLECULES DIMERIZE (JOIN TOG-
ETHER) TO FORM MOLECULES OF Al_2Cl_6. Molecules of $AlCl_3$ are HELD TOGETHER in the Al_2Cl_6
dimer BY DATIVE COVALENT BONDS.

ENERGY is RELEASED WHEN THE TWO DATIVE
BONDS ARE FORMED, and SO the DIMER IS MORE
STABLE THAN two SEPARATE $AlCl_3$ MOLECULES.

METALLIC BONDING

A METALLIC BOND IS THE ELECTROSTATIC ATTRACTION BETWEEN POSITIVE NUCLEI AND DELOCALIZED ELECTRONS.

METALLIC BONDING ONLY OCCURS BETWEEN METAL ATOMS OF the SAME ELEMENT.

The OUTERMOST ELECTRONS OF the METAL ATOMS ARE LOST AND are ABLE TO MOVE FREELY BETWEEN ATOMS, FORMING a 'SEA' OF DELOCALIZED ELECTRONS.

The METAL ATOMS are CLOSELY PACKED TOGETHER IN a REGULAR ARRANGEMENT, KNOWN AS a GIANT METALLIC LATTICE, THROUGH WHICH the 'SEA OF ELECTRONS' MOVES RANDOMLY, ATTRACTING ALL the POSITIVE NUCLEI AND BINDING THEM TOGETHER.

BELOW IS A MODEL OF METALLIC BONDING, SHOWING the OUTER SHELL ELECTRONS OF the METAL ATOMS, which MOVE RANDOMLY THROUGHOUT a LATTICE OF REGULARLY SPACED ATOMS WITH POSITIVE NUCLEI.

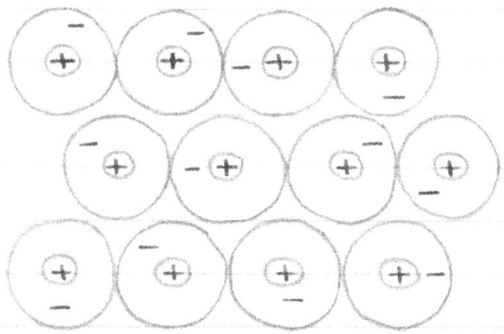

The MOBILITY OF the OUTER ELECTRONS IN METALS ACCOUNT FOR their HIGH ELECTRICAL AND THERMAL CONDUCTIVITY.

NORMALLY, the ELECTRONS MOVE RANDOMLY

THROUGHOUT the METALLIC LATTICE, BUT WHEN the METAL IS CONNECTED ACROSS A POTENTIAL DIFFERENCE, THERE IS AN OVERALL MOVEMENT OF ELECTRONS (SUPERIMPOSED ON THEIR RANDOM MOTION) FROM the NEGATIVE ELECTRODE TO the POSITIVE ELECTRODE.

The HIGH THERMAL CONDUCTIVITY OF METALS CAN also BE EXPLAINED IN TERMS OF THEIR MOB- ILE ELECTRONS. ELECTRONS IN the REGIONS OF HIGH TEMPERATURE (i.e. electrons with high kine- tic energy) MOVE RAPIDLY AND RANDOMLY TOWARDS the COOLER REGIONS OF the METAL, TRANSFERRING their ENERGY TO OTHER ELECTR- ONS THROUGHOUT the METAL.

ELECTRONEGATIVITY, POLARITY AND BONDING

ELECTRONEGATIVITY

ELECTRONEGATIVITY IS THE ABILITY OF AN ATOM IN A COMPOUND TO ATTRACT ELECTRONS IN A BOND.

THE MOST ELECTRONEGATIVE ATOMS ATTRACT BONDING ELECTRONS MOST STRONGLY.

ELECTRONEGATIVITY is MEASURED USING the PAULING ELECTRONEGATIVITY INDEX (Np), IN WHICH NUMBERS GIVE the ELECTRONEGATIVITY VALUES OF the ELEMENTS RELATIVE TO ONE ANOTHER. ELECTRONEGATIVITY VALUES are SHOWN BELOW:

H 2.1						
Li 1.0	Be 1.5	B 2.0	C 2.5	N 3.0	O 3.5	F 4.0
Na 0.9	Mg 1.2	Al 1.5	Si 1.8	P 2.1	S 2.5	Cl 3.0
K 0.8	Ca 1.0				Se 2.4	Br 2.8
					Te 2.1	I 2.5

IN GENERAL, ELECTRONEGATIVITY INCREASES MOVING FROM LEFT TO RIGHT ACROSS the PERIODIC TABLE, and DECREASES GOING DOWN A GROUP.

REACTIVE NON-METALS such as oxygen, fluorine and chlorine HAVE the HIGHEST VALUES OF ELECTRONEGATIVITY.

REACTIVE METALS such as sodium or potassium HAVE LOW VALUES OF ELECTRONEGATIVITY.

FLUORINE is the MOST ELECTRONEGATIVE ELE-
MENT and is given the arbitrary value of 4·0.

POLARITY AND BONDING

POLARIZATION IS THE DISTORTION OF CHARGE IN
A MOLECULE, RESULTING FROM UNEQUAL SHARING OF
ELECTRONS. THE MOLECULE IS SAID TO BE POLAR.

A POLAR COVALENT BOND IS ONE IN WHICH THERE
IS SEPARATION OF CHARGE BETWEEN ENDS OF THE
MOLECULE. The SEPARATION OF CHARGE IN the MOLE-
CULE IS referred to as A DIPOLE.
A POLAR COVALENT BOND is FORMED WHEN
ONE ATOM IS MORE ELECTRONEGATIVE THAN THE
OTHER.

E.G.   IN a MOLECULE OF HCl, the CHLORINE NUCLEUS
       HAS a GREATER ATTRACTION FOR the BONDING
       PAIR OF ELECTRONS THAN the HYDROGEN does,
       DUE TO ITS GREATER ELECTRONEGATIVITY.
       THIS MAKES the H ATOM SLIGHTLY POSITIVE ($\delta+$)
       AND the CHLORINE ATOM SLIGHTLY NEGATIVE ($\delta-$).

$$\overset{\delta+}{H} - \overset{\delta-}{Cl}$$

The H-Cl BOND IS POLAR AND the HCl MOLECU-
LE HAS A DIPOLE MOMENT, which is ONE OF the
SLIGHT CHARGES ($\delta+$ or $\delta-$) MULTIPLIED BY the DI-
STANCE BETWEEN THEM.

THE GREATER THE DIFFERENCE IN the ELECTR-
ONEGATIVITIES OF TWO ATOMS, THE MORE POLAR is

THE COVALENT BOND BETWEEN THEM.

POLARITY ONLY OCCURS IN NON-SYMMETRICAL MOLECULES such as HCl, WHERE THERE IS an UNEQUAL DISTRIBUTION OF CHARGE (OR WHERE the CENTRE OF POSITIVE CHARGE AND the CENTRE OF NEGATIVE CHARGE DO NOT COINCIDE).

E.G. WATER IS A POLAR MOLECULE BUT CO_2 IS NOT:

IN CARBON DIOXIDE the TWO CO BONDS are EQUALLY POLAR BUT ACT IN OPPOSITE DIRECTIONS, CANCELLING EACH OTHER OUT AND GIVING an OVERALL DIPOLE MOMENT OF ZERO.

$$\overset{\delta-}{O} = \overset{\delta+}{C} \overset{\delta+}{} = \overset{\delta-}{O}$$

$$\longleftarrow \qquad \longrightarrow$$

opposing dipole moments cancel

SINCE the DIPOLE MOMENT OF WATER IS NOT ZERO, this suggests that the WATER MOLECULE CANNOT BE LINEAR, BUT MUST BE BENT SO THAT the DIPOLES OF the TWO HO BONDS ACT IN SIMILAR DIRECTIONS AND REINFORCE EACH OTHER.

co-operating dipole moments reinforce

$$\overset{\delta+}{H} \diagdown \quad \overset{\delta-}{\ddot{O}} \diagdown \overset{\delta-}{}$$
$$\underset{\delta+}{H} \diagup$$

ALTERNATIVELY:

IN CARBON DIOXIDE, A SYMMETRICAL MOLECULE, the CENTRES OF POSITIVE AND NEGATIVE CHARGE

COINCIDE ON the CARBON ATOM, and SO the MOL-
ECULE IS NON-POLAR.

WATER HOWEVER, IS a POLAR molecule BECAUSE
the CENTRES OF POSITIVE AND NEGATIVE CHARGE
DO NOT COINCIDE. The CENTRE OF NEGATIVE CHARGE
IS LOCATED ON the OXYGEN WHILST the CENTRE OF PO-
SITIVE CHARGE IS MIDWAY BETWEEN the HYDROGEN
NUCLEI.

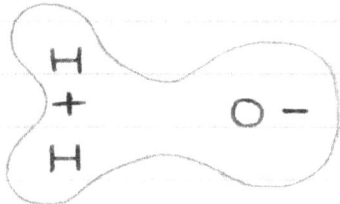

+ centre of positive charge
− centre of negative charge

E.G.   TRICHLOROMETHANE IS POLAR BUT TETRACHLORO-
        METHANE IS NOT:

IN TETRACHLOROMETHANE, EACH C−Cl BOND IS
POLAR, BUT the MOLECULE HAS a SYMMETRICAL
DISTRIBUTION OF ATOMS and SO the DIPOLES CANCEL
EACH OTHER. This RESULTS IN a NON-POLAR MOLE-
CULE.

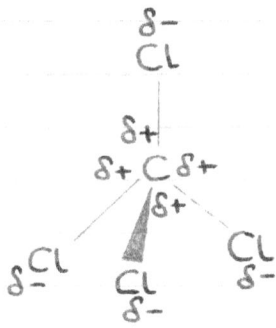

TRICHLOROMETHANE however, IS UNSYMMETRICAL,
WHICH MEANS that ITS CENTRE OF POSITIVE CHARGE

DOES NOT COINCIDE EXACTLY WITH ITS CENTRE OF
NEGATIVE CHARGE, AND a POLAR MOLECULE RESULTS.

A NON-POLAR COVALENT BOND IS ONE IN WHICH
THERE IS AN EQUAL SHARING OF THE BONDED ELE-
CTRONS.

A NON-POLAR COVALENT BOND IS FORMED BETWE-
EN TWO ATOMS WITH EQUAL OR VERY SIMILAR ELECT-
RONEGATIVITIES. IF the BONDED ATOMS ARE THE SAME,
the COVALENT BOND MUST BE NON-POLAR, as in molecules such as H_2 or Cl_2.

COVALENT BONDING USUALLY OCCURS BETWEEN
the ATOMS OF NON-METALS BECAUSE the ELECTRON-
EGATIVITY DIFFERENCES ARE SMALL, RESULTING IN
NON-POLAR OR SLIGHTLY POLAR MOLECULES.

AN IONIC BOND IS FORMED BETWEEN ATOMS
WITH A VERY LARGE DIFFERENCE IN ELECTRONEGATI-
VITY, i.e. BETWEEN a METAL ATOM AND a NON-METAL
ATOM.

THE GREATER THE DIFFERENCE IN THE ELECTRON-
EGATIVITIES OF TWO ATOMS, THE GREATER THE IONIC
CHARACTER OF THE BOND.

AS WELL AS the DISTORTION OF CHARGE DUE TO
DIFFERING ELECTRONEGATIVITIES IN COVALENT COMP-
OUNDS, THERE IS ALSO DISTORTION OF CHARGE IN IO-

NIC COMPOUNDS. THE NET POSITIVE CHARGE ON
the CATIONS CAN ATTRACT the NEGATIVE CHARGE
CLOUD OF the ANIONS WITH WHICH THEY ARE ASSO-
CIATED.

POLARIZING POWER IS the ABILITY OF a CATION,
e.g. AL^{3+} TO DISTORT the ELECTRON CLOUD OF a NE-
ARBY ANION, e.g. Cl^-.

IF the CHARGE DENSITY (AMOUNT OF POSITIVE
CHARGE IN a GIVEN VOLUME) OF the CATION INCRE-
ASES, the DISTORTION OR POLARIZATION OF the ANI-
ON INCREASES AND the COMPOUND BECOMES
LESS IONIC (MORE COVALENT) IN CHARACTER.

POLARIZING POWER OF A CATION INCREASES AS
ITS CHARGE INCREASES AND ITS SIZE DECREASES.

POLARIZABILITY OF AN ANION INCREASES AS
ITS SIZE AND CHARGE INCREASES.

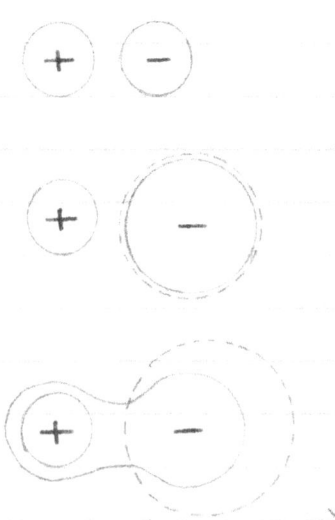

increasing polarization
of anion
by cation
(the dashed circles
represent the shapes
of the isolated anions
before polarization)

The COVALENT CHARACTER OF an IONIC COMPO-
UND IS HIGH IF the CATION IS SMALL, the ANION IS
LARGE AND the CHARGES ON the IONS ARE HIGH,

E.G. ALF_3 IS IONIC BUT $ALCl_3$ IS COVALENT. The SMA-

LL, HIGHLY CHARGED Al^{3+} CATION CANNOT POLARIZE the SMALL ELECTRON CLOUD OF F^-, BUT EASILY POLARIZES the LARGER ELECTRON CLOUD OF Cl^-.

COVALENT BONDING IS MORE PROBABLE AND IONIC BONDING LESS PROBABLE BETWEEN TWO ELEMENTS IF:

the IONS POSSESS MULTIPLE CHARGES

the ATOMS PRODUCE SMALL CATIONS OR LARGE ANIONS

INTERMOLECULAR FORCES

INTERMOLECULAR FORCES ARE the FORCES BETWEEN MOLECULES WHICH HOLD THEM TOGETHER. They are also called Van der Waals' forces. These are WEAK SHORT-RANGE FORCES ARISING FROM the ATTRACTION BETWEEN DIPOLES.

There are TWO TYPES OF INTERMOLECULAR FORCES, LONDON DISPERSION FORCES AND PERMANENT DIPOLE-PERMANENT DIPOLE ATTRACTIONS.

LONDON DISPERSION FORCES, which are more COMMONLY KNOWN AS INDUCED DIPOLE-INDUCED DIPOLE ATTRACTIONS, are ATTRACTIVE FORCES BETWEEN ALL MOLECULES.

PERMANENT DIPOLE-PERMANENT DIPOLE ATTRACTIONS, as suggested from the name, ONLY OCCUR BETWEEN POLAR MOLECULES.

INDUCED DIPOLE-INDUCED DIPOLE ATTRACTIONS

AS the ELECTRONS IN a MOLECULE ARE MOVING CONTINUALLY IN a RANDOM MOTION, at any particular moment there will be more negative charge on one side of the molecule than on the other. ONE SIDE OF the MOLECULE BECOMES SLIGHTLY POSITIVELY CHARGED ($\delta+$) AND the OTHER SIDE BECOMES SLIGHTLY NEGATIVELY CHARGED ($\delta-$), FORMING an INSTANTANEOUS DIPOLE.

This DIPOLE WILL INDUCE DIPOLES IN NEIGH-

BOURING MOLECULES BY CAUSING ELECTRONS
TO MOVE TOWARDS OR AWAY FROM ITS CHARGED
ENDS. The DIPOLES WILL ALWAYS ATTRACT BECA-
USE the ELECTRONS HAVE REDISTRIBUTED; which
means that the REPELLING CHARGED ENDS are FU-
RTHER APART THAN the ATTRACTING CHARGED
ENDS.

E.G. The FORMATION OF AN INDUCED DIPOLE in a
molecule

instantaneous   non-polar   instantaneous   induced
dipole   molecular   dipole   dipole

These INDUCED DIPOLES will ACT FIRST ONE
WAY THEN ANOTHER, CONTINUALLY OSCILLATING
as a result of electron movement:

$$\delta - \quad \delta + \quad \longleftrightarrow \quad \delta + \quad \delta -$$

PROVIDED that the MOLECULES REMAIN CLOSE
TO EACH OTHER, the DIPOLES will OSCILLATE IN
SYNCHRONIZATION so that the attraction is maintained.
This synchronized oscillation of dipoles can occur over
large numbers of molecules as long as they are close
enough together.

Strength of induced dipole - induced dipole attractions

The STRENGTH OF INDUCED DIPOLES INCR-
EASES WITH an INCREASE IN the NUMBER OF E-
LECTRONS BECAUSE INDUCED DIPOLES ARE LARGER,

SO GENERALLY CLOSER TOGETHER, AND CAN DEVELOP OVER a GREATER DISTANCE.

E.G. The boiling points of the Noble Gases increases going down the group.

helium	− 269°C
neon	− 246°C
argon	− 186°C
Krypton	− 152°C
Xenon	− 107°C
radon	− 62°C

The NOBLE GASES are MONATOMIC MOLECULES which exist in the gaseous state at room temperature.

In the liquid state, ENERGY is REQUIRED TO BOIL the NOBLE GASES, and the energy required INCREASES ON DESCENDING the GROUP, providing EVIDENCE FOR INDUCED DIPOLE − INDUCED DIPOLE FORCES BETWEEN MOLECULES, which increase with additional shells of electrons.

The SHAPES OF MOLECULES also AFFECT the STRENGTH OF INDUCED DIPOLE − INDUCED DIPOLE ATTRACTIONS.

LINEAR MOLECULES DEVELOP LARGER INSTANTANEOUS DIPOLES due to greater electron movement THAN GLOBULAR MOLECULES containing the same number of electrons.

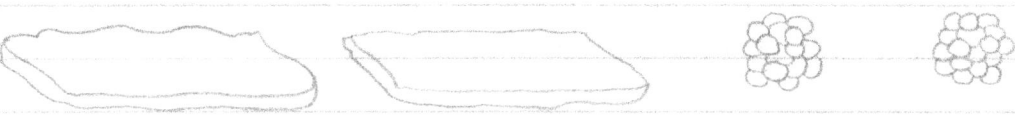

linear molecules   globular molecules

Also, there are LESS INDUCED DIPOLES IN GLOBULAR MOLECULES DUE TO FEWER POINTS OF CONTACT THAN in LINEAR MOLECULES, where long molecular chains lie close and relatively parallel to one another.

E.G. butane has a higher boiling point than 2-methyl propane

butane $\qquad CH_3 - CH_2 - CH_2 - CH_3 \qquad -0.5°C$

2-methyl propane $\qquad CH_3 - CH - CH_3 \qquad -11.7°C$
$\qquad\qquad\qquad\qquad\qquad\quad | $
$\qquad\qquad\qquad\qquad\qquad\ CH_3$

Butane has a higher boiling point than 2-methyl propane because the induced dipoles are larger and the straight chain molecules lie closer together than the branched chain 2-methyl propane molecules.

PERMANENT DIPOLE - PERMANENT DIPOLE ATTRACTIONS

MOLECULES THAT ARE POLAR are said to HAVE A PERMANENT DIPOLE.
PERMANENT DIPOLE-PERMANENT DIPOLE ATTRACTIONS ARE THE ATTRACTIONS BETWEEN THE POSITIVE END OF ONE POLAR MOLECULE AND THE NEGATIVE END OF ANOTHER POLAR MOLECULE.

E.G. permanent dipole - permanent dipole attraction between HCl molecules

$$\overset{\delta+}{H} - \overset{\delta-}{Cl} --- \overset{\delta+}{H} - \overset{\delta-}{Cl} --- \overset{\delta+}{H} - \overset{\delta-}{Cl}$$

PERMANENT DIPOLE—PERMANENT DIPOLE ATTRACTIONS OCCUR IN ADDITION TO LONDON DISPERSION FORCES between polar molecules, and although both are weak, the PERMANENT DIPOLE—PERMANENT DIPOLE INTERACTIONS ARE MUCH STRONGER THAN the DISPERSION FORCES.

Therefore SUBSTANCES which contain MOLECULES WITH PERMANENT DIPOLES tend to have HIGHER MELTING AND BOILING POINTS THAN those consisting of NON—POLAR MOLECULES.

E.G. propanone (CH_3COCH_3) is a liquid but butane ($CH_3CH_2CH_2CH_3$) is a gas.

DUE TO the ELECTRONEGATIVE OXYGEN ATOM, AND the SHAPE OF the MOLECULE, PROPANONE HAS an OVERALL PERMANENT DIPOLE. SINCE BOTH PROPANONE AND BUTANE HAVE SIMILAR INDUCED DIPOLE—INDUCED DIPOLE ATTRACTIONS (same number of electrons), the HIGHER BOILING POINT OF PROPANONE must be DUE TO the ADDITIONAL PERMANENT DIPOLE—PERMANENT DIPOLE ATTRACTIONS BETWEEN its MOLECULES, which REQUIRE EXTRA ENERGY TO BE BROKEN.

HYDROGEN BONDING

The graph below shows the boiling points of the hydrides in groups 4, 5, 6 and 7.

The BOILING POINTS OF the GROUP 4 HYDRIDES INCREASE AS the MOLECULES GET LARGER from CH_4 to SnH_4 BECAUSE the NUMBER OF ELECTRONS INCREASE and so the London dispersion forces are greater.

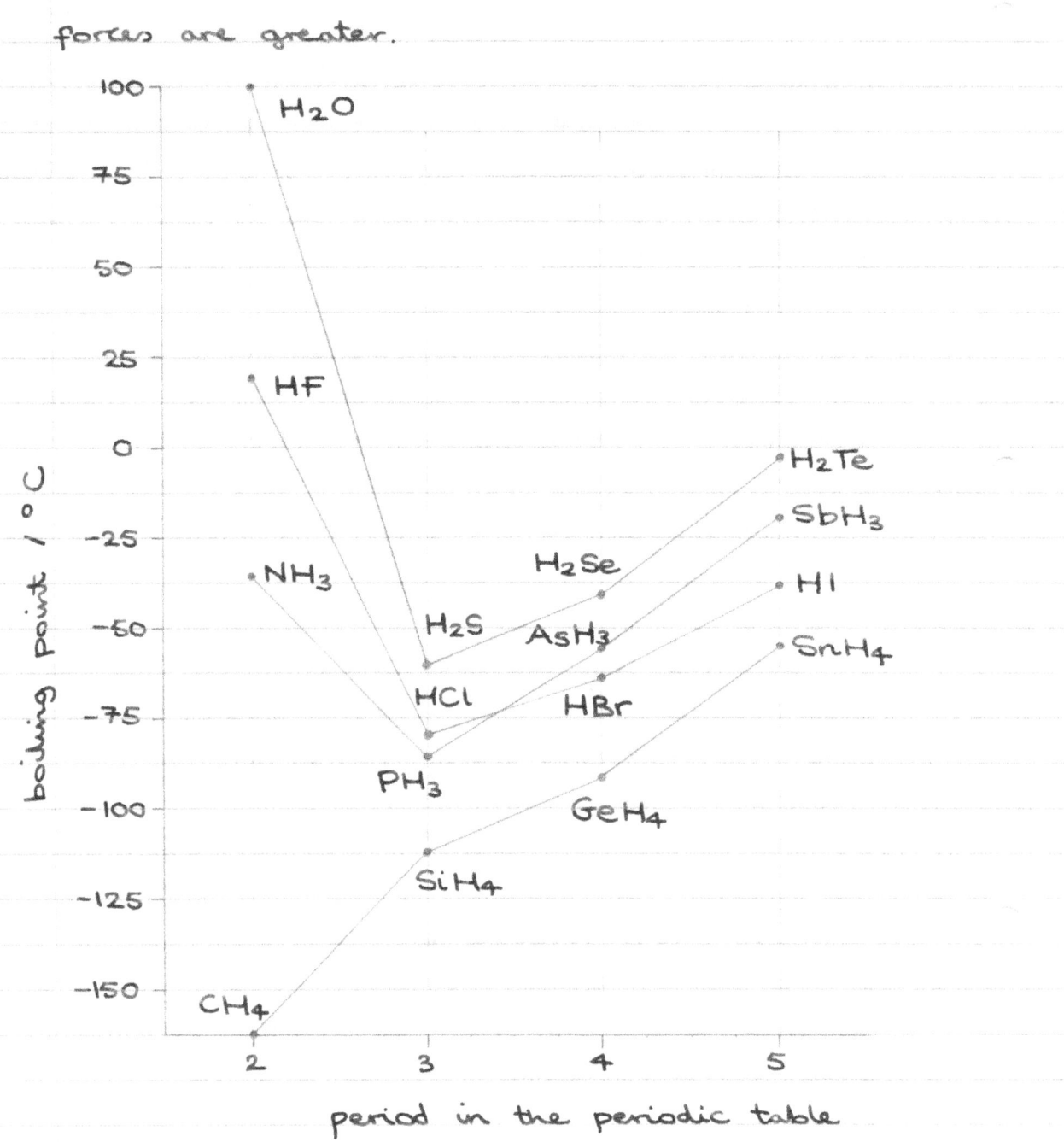

period in the periodic table

A SIMILAR TREND IN BOILING POINTS OCCURS
FOR the HYDRIDES OF GROUPS 5, 6 AND 7, EXCE-
PTING the HYDRIDE OF the FIRST ELEMENT IN
EACH GROUP, WHOSE BOILING POINT IS UNUSU-
ALLY HIGH.

H_2O, NH_3 AND HF are all VERY POLAR BEC-
AUSE THEY CONTAIN the THREE MOST ELECTRON-

EGATIVE ELEMENTS (OXYGEN, NITROGEN AND FL-
UORINE) LINKED DIRECTLY TO HYDROGEN, which is
weakly electronegative. This RESULTS IN EXCEPTIO-
NALLY POLAR MOLECULES WITH STRONGER INT-
ERMOLECULAR FORCES THAN USUAL, known as
hydrogen bonds.

A HYDROGEN BOND IS A WEAK BOND BETW-
EEN THE LONE PAIR ON A VERY ELECTRONEG-
ATIVE ATOM (N, O OR F) AND A HYDROGEN AT-
OM COVALENTLY BONDED TO A VERY ELECTRON-
EGATIVE ATOM.
The HIGHLY ELECTRONEGATIVE N, O or F ATOM
POLARIZES the COVALENT BOND WITH HYDROGEN
SO much that the NUCLEUS OF the HYDROGEN
ATOM IS EXPOSED TO ATTRACTION BY a LONE
PAIR ON ANOTHER ELECTRONEGATIVE ATOM.
HYDROGEN BONDING is STRONGER THAN
LONDON DISPERSION FORCES AND PERMANENT
DIPOLE- PERMANENT DIPOLE ATTRACTIONS BUT
WEAKER THAN COVALENT BONDING. HYDROGEN
BONDS are about 5-10% as strong as covalent
bonds and are EASILY DISRUPTED BY PHYSICAL
METHODS, e.g. the SURFACE TENSION OF WATER
IS LOWERED BY ADDING a SOAP BECAUSE OF a
REDUCTION IN the NUMBER OF HYDROGEN BONDS
BETWEEN WATER MOLECULES.
Shown below is hydrogen bonding between
molecules of :

AMMONIA

WATER

$\delta+$ $\overset{..}{\underset{..}{\delta-}}$ $\delta+$ $\overset{..}{\underset{..}{\delta-}}$ $\delta+$ $\overset{..}{\underset{..}{\delta-}}$

- - - H — O: - - - H — O: - - - H — O: - - -
        | | |
        H $\delta+$ H $\delta+$ H $\delta+$

HYDROGEN
FLUORIDE

$\delta+$ $\delta-$ $\delta+$ $\delta-$ $\delta+$ $\delta-$

- - - H — F: - - - H — F: - - - H — F: - - -

    Looking at the graph again, we can see
that WATER HAS a HIGHER BOILING POINT THAN
HYDROGEN FLUORIDE AND AMMONIA. This is BEC-
AUSE the AVERAGE NUMBER OF HYDROGEN BON-
DS BETWEEN a PAIR OF WATER MOLECULES IS
TWO (as there are two O—H bonds and two lone-
pairs), WHEREAS IT IS ONLY ONE BETWEEN a
PAIR OF HYDROGEN FLUORIDE OR AMMONIA
MOLECULES.

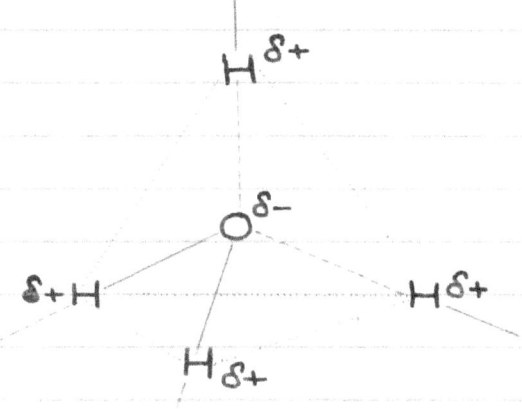

SHAPES OF MOLECULES AND IONS

The SHAPES OF MOLECULES AND IONS CAN BE PREDICTED USING the ELECTRON PAIR REPULSION THEORY.

The SHAPE OF a MOLECULE OR ION DEPENDS UPON the NUMBER OF (σ BOND and LONE) ELECTRON PAIRS IN the OUTER SHELL OF the CENTRAL ATOM.

THE ELECTRON PAIRS POSITION THEMSELVES AS FAR APART AS POSSIBLE SO THAT REPULSION BETWEEN THEM IS MINIMIZED.

The ORBITAL OCCUPIED BY a LONE PAIR OF ELECTRONS IS SMALLER AND CLOSER TO the NUCLEUS OF an ATOM THAN a BONDED PAIR. This means that a LONE PAIR (Lp) CAN EXERT a GREATER REPELLING EFFECT THAN a BONDED PAIR (bp).

LONE PAIR — LONE PAIR REPULSION is GREATER THAN LONE PAIR — BOND PAIR REPULSION, which is GREATER THAN BOND PAIR — BOND PAIR REPULSION.

$$lp - lp > lp - bp > bp - bp$$

IN MOLECULES OR IONS CONTAINING DOUBLE OR TRIPLE BONDS, the DOUBLE OR TRIPLE BOND COUNTS AS a SINGLE BONDING PAIR OF ELECTRONS (the σ BOND ALONE is CONSIDERED), SINCE the ELECTRON PAIR IN A π BOND ONLY HAS A SLIGHT EFFECT ON the SHAPE OF a MOLECULE OR ION.

WHEN DRAWING the SHAPE OF a MOLECULE OR ION, BONDS IN the PLANE OF the PAPER are drawn

—— , BONDS POINTING BEHIND the PLANE OF the PAPER are drawn ----, and BONDS COMING OUT OF the PLANE OF the PAPER are drawn ◄——.

The SHAPES AND BOND ANGLES GIVEN BY DIFFERENT NUMBERS OF (σ BOND) ELECTRON PAIRS AROUND a CENTRAL ATOM ARE SHOWN BELOW.

NO. OF σ bond ELECTRON PAIRS	BOND ANGLE (°)	SHAPE	DIAGRAM
2	180	linear	
3	120	trigonal planar	
4	109.5	tetrahedral	
5	120, 90	trigonal bipyramidal	
6	90	octahedral	

EACH LONE PAIR OF ELECTRONS IN the SAME PLANE AS the BONDING PAIRS WILL DECREASE the BOND ANGLE BY 2.5° DUE TO THEIR GREATER REPELLING EFFECT.

EXAMPLES

NO. OF ELECTRON PAIRS AROUND CENTRAL ATOM ✳	NO. OF LONE PAIRS	SHAPE	BOND ANGLES (°)	EXAMPLES		
2	0	LINEAR	180	$Cl-Be-Cl$ $BeCl_2$	$O=C=O$ CO_2	$H-C\equiv N$ HCN
3	0	TRIGONAL PLANAR	120	CO_3^{2-}	NO_3^-	BCl_3
3	1	BENT OR V-SHAPED	—	SO_2 (119.5°)		NO_2^- (115°)
4	0	TETRA-HEDRAL	109.5	CH_4	NH_4^+	SO_4^{2-}
4	1	PYRAM-IDAL	107	NH_3		SO_3^{2-}

NO. OF ELECTRON PAIRS AROUND CENTRAL ATOM*	NO. OF LONE PAIRS	SHAPE	BOND ANGLES (°)	EXAMPLES
4	2	BENT OR V-SHAPED	104.5	H_2O H_2S
5	0	TRIGONAL BIPYRA-MIDAL	120, 90	PCl_5
6	0	OCTAH-EDRAL	90	SF_6
6	2	SQUARE PLANAR	90	XeF_4

* EXCEPTING THOSE IN π BONDS

HYDRATED COMPLEX IONS

A COMPLEX ION IS A CENTRAL METAL CATION SURROUNDED BY ANIONS OR MOLECULES CALLED LIGANDS. EACH LIGAND CONTAINS AT LEAST ONE ATOM WITH A LONE PAIR OF ELECTRONS. These CAN BE DONATED TO the CENTRAL CATION FORMING a CO-ORDINATE BOND. The LIGAND is said to be CO-ORDINATED TO the CENTRAL ION.

The NUMBER OF LIGANDS BONDED TO the CENTRAL ION IS CALLED THE CO-ORDINATION NUMBER and MAY BE 2,4 (fairly common) OR 6 (most common).

WATER is by far the MOST COMMON LIGAND. IN AQUEOUS SOLUTION, METAL CATIONS EXIST AS HYDRATED COMPLEXES WITH WATER MOLECULES. (HYDRATION IS THE ATTACHMENT OF WATER MOLECULES TO IONS IN AQUEOUS SOLUTION).

WHEN MAGNESIUM SULPHATE DISSOLVES IN WATER, IONS are PRODUCED. SIX WATER MOLECULES are ATTACHED TO the MAGNESIUM ION BY DATIVE COVALENT BONDS FORMED BY the OXYGEN ATOM DONATING a LONE ELECTRON PAIR.

Thus, an Mg^{2+} (aq) ion is represented as $[Mg(H_2O)_6]^{2+}$ (aq) which is called a HEXAAQUAMAGNESIUM ION:

Similarly, ALUMINIUM CHLORIDE, although covalent, CAN DISSOLVE IN WATER TO PRODUCE IONS.
SIX WATER MOLECULES BOND TO the ALUMINIUM
ION, GIVING a HEXAAQUAALUMINIUM ION,
$[Al(H_2O)_6]^{3+}$ (aq):

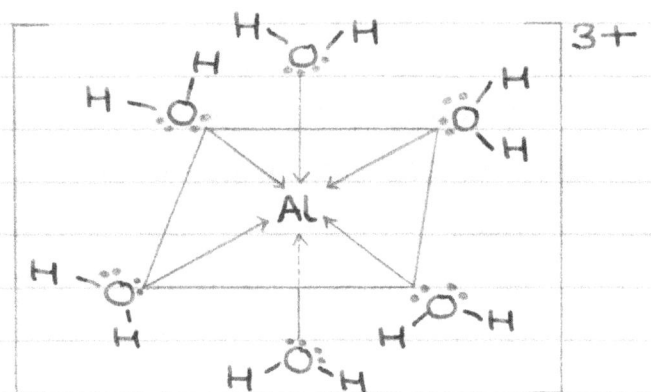

IN BOTH THE HYDRATED MAGNESIUM AND
ALUMINIUM IONS, the LIGANDS OCCUPY OCTAHEDRAL
POSITIONS, since both complexes have a co-ordinat-
ion number of SIX. (THE SIX ELECTRON PAIRS
AROUND the CENTRAL METAL ION are REPELLED
AS FAR AS POSSIBLE FROM EACH OTHER).

STRUCTURE

IONIC STRUCTURES

IONIC COMPOUNDS FORM GIANT (endlessly rep-
eating) LATTICES OF OPPOSITELY CHARGED IONS.
The ions are HELD TOGETHER BY ELECTROSTATIC
ATTRACTION.

Different ionic structures have their ions arranged
in different patterns. One of the simplest arrangements
is the sodium chloride structure.

Sodium Chloride

In SODIUM CHLORIDE, the Na$^+$ and Cl$^-$ IONS
are ARRANGED IN a CUBIC LATTICE as shown
below.

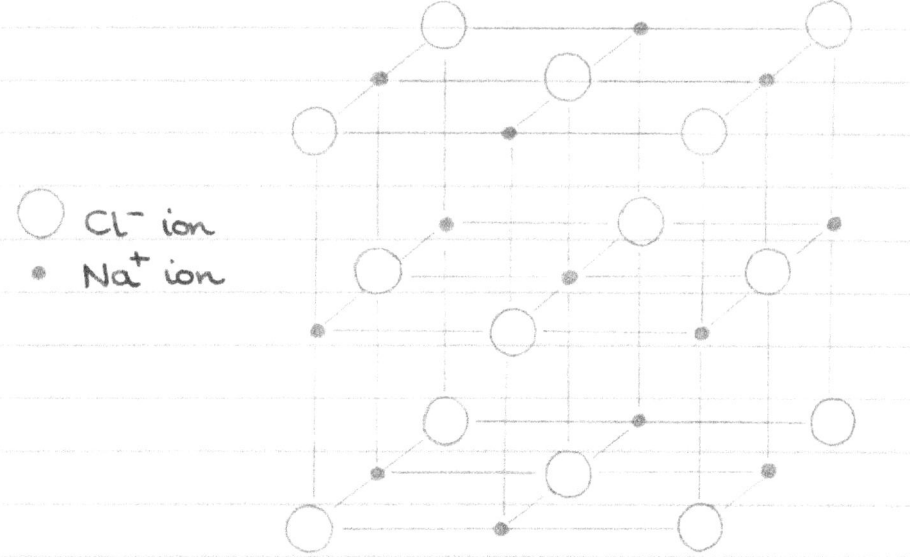

○ Cl$^-$ ion
• Na$^+$ ion

The SODIUM ATOMS FILL the HOLES IN the
LATTICE OF CHLORIDE IONS.
IN AN IONIC LATTICE, THE CO-ORDINATION
NUMBER OF AN ION IS THE NUMBER OF NEARE-

ST, EQUIDISTANT, OPPOSITELY CHARGED IONS.

In the sodium chloride lattice each positive sodium ion is surrounded by six Cl^- ions and each negative chloride ion is surrounded by six Na^+ ions. The STRUCTURE OF SODIUM CHLORIDE is SAID TO HAVE 6:6 CO-ORDINATION because the Na^+ ions have a co-ordination number of 6 and the Cl^- ions also have a co-ordination number of 6.

The ARRANGEMENT OF IONS IN the SODIUM CHLORIDE LATTICE is DESCRIBED AS a FACE-CENTRED CUBIC BECAUSE the IONS ARE ARRANGED IN A CUBE WITH THE SAME IONS AT EACH CORNER AND AT THE CENTRE OF EACH FACE.

Caesium Chloride

Another simple ionic structure is that shown by caesium chloride. The general shape of the lattice is again cubic, but is different in detail from sodium chloride:

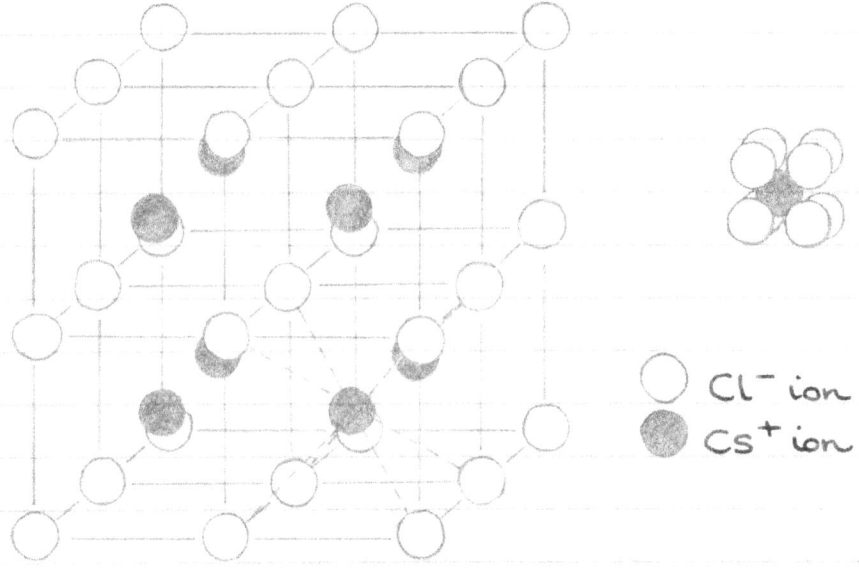

○ Cl^- ion
● Cs^+ ion

Unlike Na^+, the Cs^+ ION IS SIMILAR IN SIZE

TO the Cl^- ION, WHICH MEANS THAT THE Cs^+ IONS
ARE TOO LARGE TO FILL the HOLES IN the LAT-
TICE OF CHLORIDE IONS. INSTEAD, the CHLORIDE
IONS ADOPT A MORE OPEN (SIMPLE) CUBIC ARR-
ANGEMENT, WITH the Cs^+ IONS ENTERING the
GAPS BETWEEN THEM. THE TWO LATTICES ARE
INTERPENETRATING, IN AN ARRANGEMENT CALL-
ED A DOUBLE SIMPLE CUBIC. Since each Cs^+ ion
has eight Cl^- ions around it and vice versa, the
STRUCTURE OF CAESIUM CHLORIDE is SAID TO
HAVE 8:8 CO-ORDINATION.

Properties of ionic compounds

1) HIGH MELTING TEMPERATURE

OPPOSITELY CHARGED IONS are HELD TOGET-
HER in the lattice BY STRONG ELECTROSTATIC FOR-
CES, WHICH NEED A LARGE AMOUNT OF ENERGY
(called the lattice energy) TO OVERCOME. Therefore
ionic compounds have high melting points.

2) HARD AND BRITTLE

EVERY ION in an ionic solid IS HELD IN the
lattice BY STRONG ATTRACTIONS FROM the OPPOS-
ITELY CHARGED IONS AROUND IT, which means IO-
NIC SOLIDS ARE HARD AND DIFFICULT TO CUT.
However, STRESSES CAN DISPLACE ONE LAYER
OF IONS RELATIVE TO THE NEXT, CAUSING IONS
OF the SAME CHARGE TO COME TOGETHER. This
CAUSES BRITTLENESS, AS REPULSION OCCURS,
forcing apart the two portions of the lattice.

arrangement of one
layer of ions before
displacement

arrangement of ions
after displacement

3) ELECTROLYTES

IN the SOLID LATTICE the IONS ARE IN FIXED
POSITIONS AND THERE ARE NO MOBILE CHARGE
CARRIERS, therefore the ionic compound is a non-co
nductor.

WHEN MOLTEN OR IN AQUEOUS SOLUTION the
IONS ARE FREE TO MOVE and so conduct electricity.

GIANT ATOMIC STRUCTURES

ELEMENTS SUCH AS CARBON AND SILICON,
which have medium electronegativity values and can
make four covalent bonds FORM GIANT ATOMIC ST-
RUCTURES OF COVALENTLY BONDED ATOMS.

DIAMOND AND GRAPHITE ARE BOTH ALLOTR-
OPES OF CARBON THAT FORM GIANT ATOMIC STRU-
CTURES. (ALLOTROPES ARE DIFFERENT FORMS OF
THE SAME ELEMENT IN THE SAME STATE).

Diamond

IN DIAMOND, EVERY CARBON ATOM IS AT the CENTRE OF a REGULAR TETRAHEDRON, SURROUNDED AT the CORNERS BY FOUR OTHER CARBON ATOMS to which it is covalently bonded.

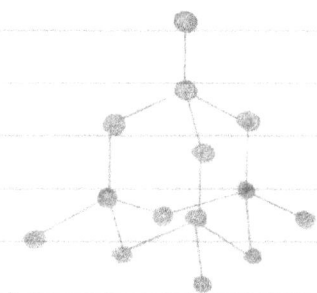

Properties of diamond

1) HIGH MELTING TEMPERATURE

The LATTICE is HELD TOGETHER BY MANY STRONG CARBON-CARBON COVALENT BONDS which need to be broken before melting occurs.

2) HARD

IN ORDER TO DISTORT the CRYSTAL, MANY STRONG COVALENT BONDS NEED TO BE BROKEN THROUGHOUT the 3-dimensional LATTICE.
Also, the TETRAHEDRAL SHAPE ENABLES EXTERNAL FORCES TO BE SPREAD OUT THROUGHOUT the LATTICE.

3) NON-CONDUCTOR OF ELECTRICITY

ELECTRONS ARE LOCALIZED IN COVALENT
BONDS between carbon atoms, SO CANNOT MOVE.

Graphite

IN GRAPHITE EACH CARBON ATOM is COVAL-
ENTLY BONDED TO THREE OTHERS IN LARGE FLAT
SHEETS OF INTERCONNECTED HEXAGONS, which are
ARRANGED INTO PARALLEL LAYERS. The distance
between layers is 0.335 nm, and the FOURTH ELE-
CTRON is DELOCALIZED OVER the WHOLE LAY-
ER. Weak dispersion forces hold the layers together.

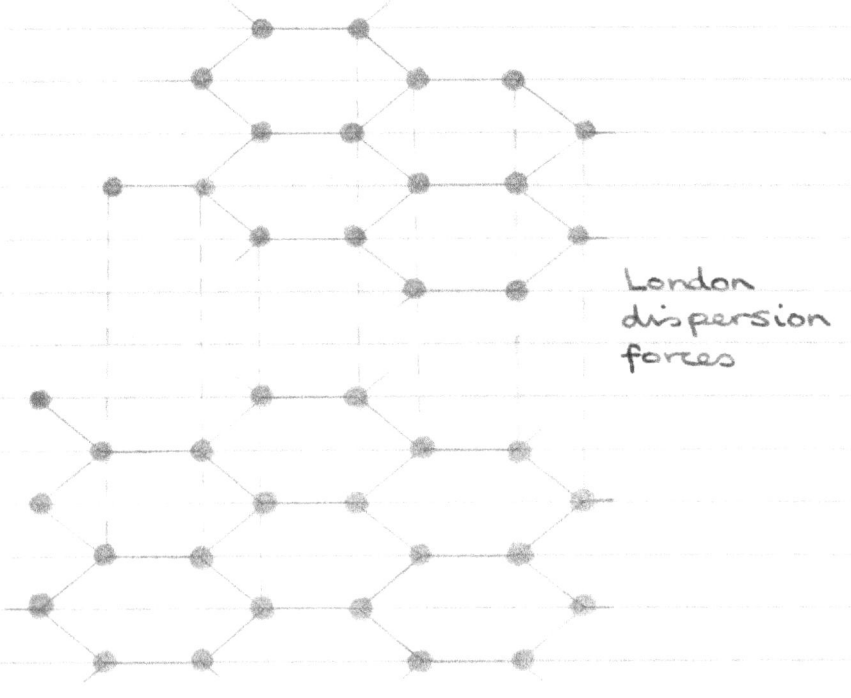

London
dispersion
forces

Properties of graphite

1) HIGH MELTING TEMPERATURE

STRONG COVALENT BONDS WITHIN LAYERS
need to be broken in order to melt graphite.

2) SOFT

WEAK DISPERSION FORCES BETWEEN LAY-ERS ALLOW THEM TO SLIDE OVER EACH OTHER EASILY. This gives graphite its slippery lubricating property.

3) CONDUCTOR OF ELECTRICITY

Within each sheet SOME OF THE VALENCE ELECTRONS (not used in bonding) ARE FREE TO MOVE. These mobile electrons give graphite its good electrical conductivity.

SIMPLE MOLECULAR STRUCTURES

SIMPLE MOLECULAR SUBSTANCES FORM CR-YSTAL LATTICES IN WHICH MOLECULES ARE HELD TOGETHER BY HYDROGEN BONDS OR VAN DER WAALS' FORCES.

IODINE

The ARRANGEMENT OF IODINE MOLECULES IN the CRYSTAL LATTICE is DESCRIBED AS a FACE-CENTRED CUBIC.

THERE ARE WEAK LONDON DISPERSION FOR-CES BETWEEN MOLECULES holding them together.

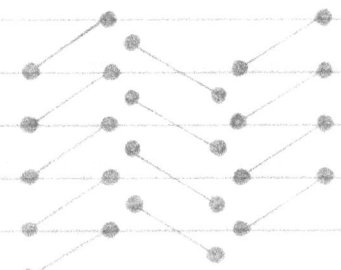

Ice

The ARRANGEMENT OF WATER MOLECULES IN ICE CREATES a VERY OPEN TETRAHEDRAL STRUCTURE. EACH OXYGEN ATOM in ice is SURROUNDED TETRAHEDRALLY BY FOUR OTHERS, AN ARRANGEMENT RESULTING FROM the PRESENCE OF TWO HYDROGEN ATOMS AND TWO LONE ELECTRON PAIRS IN EACH WATER MOLECULE.

Water has higher than expected melting and boiling points due to hydrogen bonding between molecules. HYDROGEN BONDS are CONSTANTLY BEING MADE AND BROKEN AS WATER MOLECULES MOVE PAST EACH OTHER. AS WATER IS COOLED TO 0°C, the MOLECULES MOVE MORE SLOWLY AND the HYDROGEN BONDING BECOMES PERMANENT, linking oxygen atoms together in a tetrahedral arrangement.

The open structure of ICE means that it IS LESS DENSE THAN WATER AT 0°C (ice floats on water). WHEN ICE MELTS THIS STRUCTURE BREAKS UP AND the WATER MOLECULES APPROACH EACH

OTHER MORE CLOSELY, SO that the LIQUID HAS A HIGHER DENSITY.

Polymers

IN A LARGE POLYMER with many contacts, the LONDON DISPERSION FORCES between molecules CAN BE APPRECIABLE.

These dispersion forces MAKE a SIGNIFICANT CONTRIBUTION TO the STRENGTH OF a NON-POLAR POLYMER such as polyethene.

Experiments have shown that the TENSILE STRENGTH OF HIGH-DENSITY POLYETHENE, which has tightly packed parallel molecules, IS THREE TIMES AS LARGE AS that of LOW-DENSITY POLYETHENE, which is packed less tightly and therefore has weaker dispersion forces.

Properties of molecular substances

1) LOW MELTING TEMPERATURE

Molecular substances are held together by WEAK VAN DER WAALS' FORCES which are EASILY OVERCOME AT LOW TEMPERATURES. The presence of hydrogen bonding will increase the melting points.

2) SOFT

Weak intermolecular forces allow MOLECULES to be SEPARATED EASILY, hence crystals of molecular substances are usually soft.

3) NON-CONDUCTORS OF ELECTRICITY

MOLECULAR COMPOUNDS CONTAIN NEITHER DELOCALIZED ELECTRONS (like metals) NOR IONS (like ionic compounds). Thus, they cannot conduct electricity even when molten or dissolved in water.